Modeling Complex Computer and Communication Systems: A Domain-Oriented Design Framework

Other Books in the McGraw-Hill Series on Computer Communications

BATES • *Voice and Data Communications Handbook*

BENNER • *Fibre Channel*

BLACK • *Frame Relay Networks, 2/e*

ENCK AND BECKMAN • *LAN to WAN Interconnection*

GEIER • *Network Reengineering*

GORALSKI • *Introduction to ATM Networking*

GRINBERG • *Computer/Telecom Integration*

HELDMAN • *The Telecommunications Information Millenium*

HERTZOFF • *AT&T Global Messaging*

KUMAR • *Broadband Communications*

LEE • *Mobile Cellular Telecommunications, 2/e*

MINOLI • *Video Dialtone Technology*

PECAR, O'CONOR, AND GARBIN • *The McGraw-Hill Telecommunications Factbook*

RODDY • *Satellite Communications, 2/e*

RUSSELL • *Signaling System #7*

Modeling Complex Computer and Communication Systems: A Domain-Oriented Design Framework

Oryal Tanir
Bell Canada
Quebec, Canada

McGraw-Hill
New York San Francisco Washington, D.C. Auckland Bogotá
Caracas Lisbon London Madrid Mexico City Milan
Montreal New Delhi San Juan Singapore
Sydney Tokyo Toronto

Library of Congress Cataloging-in-Publication Data

Tanir, Oryal.
Modeling complex computer and communication systems : a domain-oriented design framework / Oryal Tanir.
p. cm.
Includes bibliographical references and index.
ISBN 0-07-063312-6
1. Telecommunication systems—Design. 2. Telecommunication systems—Simulation methods. 3. System design. 4. Computer networks—Design. I. Title.
TK5102.5.T318 1996
621.382—dc20 96-9381
CIP

McGraw-Hill

A Division of The ***McGraw-Hill*** *Companies*

1 2 3 4 5 6 7 8 9 0 DOC/DOC 9 0 1 0 9 8 7 6

ISBN 0-07-063312-6

The sponsoring editor for this book was John Wyzalek, the editing supervisor was Bernard Onken, and the production supervisor was Suzanne W. B. Rapcavage. It was set in Century Schoolbook by Estelita F. Green of McGraw-Hill's Professional Book Group composition unit.

Printed and bound by R. R. Donnelley & Sons Company.

This book is printed on recycled, acid-free paper containing a minimum of 50% recycled, de-inked fiber.

Contents

Preface

This book is about the design of complex communication systems. In particular, it advocates the use of modeling and simulation as early as possible in the development process. To this end, the book provides the foundation and framework for constructing complex systems, from the inception of an idea until the synthesis of the end product. The material covered in the book touches upon many disciplines and attempts to bind many of the existing concepts through an intermediate model representation to form a uniform framework for design.

The reader will notice a strong emphasis on the application of the concepts through examples derived from the telecommunications domain. In essence, the framework that is developed is domain dependent, which is the reason for using strictly telecommunications examples. However, the concepts are applicable to any system which can be modeled using discrete event representations.

The need to write this book arose from the lack of a cohesive integration of ideas among the various disciplines that contribute to the design of a product. When viewing the design literature, one can find books on an assortment of specialized areas, such as software systems design (object-oriented languages and processes), hardware systems design (hardware description languages), specialized tools (computer aided design frameworks, computer-aided system engineering), analysis techniques (formal languages, specification, simulation), and an assortment of processes. However, there are not many that address the issues confronted by designers during the conceptual stages of the design. At this stage, the decisions made will affect both the hardware and software realization of the end product. A unified view of the system, capturing both the hardware and software elements, can facilitate a more pragmatic exploration of design alternatives based upon the designer's experience as well as the knowledge inherent in the design environment. This book attempts to address these issues.

Many of the concepts that are introduced are a result of five years of research and collaboration between the author, Vinod K. Agarwal (Logic Vision), and Pramod C. P. Bhatt (Indian Institute of Technology) as part of a Bell Canada and McGill University research initiative. Also, many of the examples and problems that are presented are derived from real problems encountered by the author as a design consultant (design modeling and simulation of telecommunications systems) over the last ten years.

Because of the broad scope of the book, many different types of readers may find it useful:

- *System designers* will appreciate the inclusion of examples and case studies to complement the concepts that are introduced.
- *Development and project managers* will understand the role that modeling and simulation can play in early design, their benefits, and their realization.
- *Hardware and software engineers* can benefit from the modeling and synthesis applications.
- *Researchers and practitioners* in the areas of performance modeling, simulation, specification, and high-level synthesis may also find the material relevant to their interests.
- *Computer design students* can use the book as a source book to navigate through interrelated concepts and disciplines in design.

The material in this book becomes progressively more detailed. The first three chapters serve as an introduction and background to design and modeling concepts. Chapters 4 to 6 present the concepts necessary for the construction of a design framework capable of modeling at highly abstract levels. Chapter 7 presents an implementation of the framework. Chapter 8 presents detailed case studies using the models developed. Finally, Chapter 9 examines some of the more advanced issues. The appendices provide source code listing for parts of the design framework developed.

Oryal Tanir

Acknowledgments

This book would not have been possible without the support and guidance of many different individuals. Professors Vinod K. Agarwal and Pramod C. P. Bhatt have helped to mold and inspire many of the concepts that resulted in the Design Analysis and Synthesis Environment. I would also like to acknowledge Dr. Hakan Erdogmus for his constructive criticism of the draft manuscript, which has resulted in a more legible document. I would also like to acknowledge the support of Bell Canada during the preparation of the book.

On another level, no less important, I am also indebted to my wife, children, and parents for their endless support and patience during the creation of this work.

Oryal Tanir

Chapter

1

Introduction

The telecommunication network is arguably the largest man-made machine in the world, composed of an elaborate blend of hardware and software elements. As in other high-technology-driven fields, market pressures are forcing the telecommunication industry to produce higher-quality products in shorter times. Since its inception in the late nineteenth century, the technology has evolved from telegraph lines to fiber and wireless communication systems. Service expectations for such systems have evolved from the traditional voice traffic to include data and video.

Like other high-technology systems, the typical telecommunication network is realized with sophisticated digital elements such as switches, transport facilities, and fiber transmission equipment. Developments in the design of digital systems have been ongoing in the last few decades. Early designs were constructed using discrete devices and were limited to a few dozen transistors. Today a reasonable system will contain dozens of microchips, each with possibly several million transistors. Although the earlier designs could be attempted with just a paper and pencil, the design of modern systems is impossible without extensive computer assistance. As the state of the technology evolves and more complex systems are contemplated, even more sophisticated tools will be required.

A key ingredient for success in designing complex systems is an environment that can model and represent a designer's conceptual system in some computer-readable form. The model must be at a sufficiently abstract level to permit the designer to concentrate on relevant details and not wallow in unnecessary detail. The designer's emphasis must be more on model manipulation and exploration of solutions to the problem than on the process of creating the model.

This book deals with the fundamental issues of modeling and design of complex systems. It presents the concepts needed to capture complex designs and a framework for implementing them. In accomplishing its goal, the book touches upon many different aspects of the design process. In particular, the reader will find examples that are related to the telecommunications domain, but are equally applicable to other computer-based systems.

Perspective

From a historical perspective, the communication age is rich with technological breakthroughs. Some examples are the discovery of the transistor, the introduction of digital transmission, and the proliferation of personal computers. Today, we are undergoing another significant communications revolution whose implications may be felt in the upcoming generations of products and services. This is reflected in the increased demand for access to electronic data across the public and private networks. Traditionally, public and private communication needs have been treated as two separate entities. In the former, digital switches dominate the market, providing constant-rate services (such as voice traffic) to a very large user base. In the latter, variable-bit-rate traffic needs are supported through local area networks composed mostly of workstations, routers, and bridges.

The last few years have resulted in a paradigm shift in the desktop computer domain. With the proliferation of personal computers and inexpensive communication facilities for users through Internet service providers, the relationship between users and the communication network has been changing radically. Users are feeling the limitations of current modem access speeds. As data requirements have increased, the existing networks have reached a technological bottleneck. Private networks deployed in businesses are demanding more and more distributed service across geographically wider areas. The voice switches are inefficient at transporting large quantities of data. Overlay networks (such as frame relay) which parallel the digital voice network have been put in place to overcome some of the data-transport difficulties. However, these are inefficient for smaller, bursty data traffic. Furthermore, anticipated demands such as those for multimedia and video applications cannot effectively be supported in a cost-efficient manner by any of the existing overlay networks.

To fulfill the technical demands of bridging the private and public networks, broadband components such as asynchronous transfer mode (ATM) have been developed that promise to answer most of the technical issues. With the deployment of broadband systems, the supported bandwidth between users is becoming larger. Computing in the traditional sense is no longer perceived as an isolated activity.

Networking is an important aspect and will certainly be a technology driver for the near future. Today's computer is a processing node interconnected through a *web* to other processing nodes, each running applications supported by their web browsers. The underlying framework supporting the communication (akin to a distributed system's hardware) is the telecommunication network.

Design Complexity

The design of traditional specialized products (such as telephone switches) to be deployed in the network is not unlike the design of a computer system. The modern digital switch is an example of the level of complexity that can be encountered in a network. Such a device incorporates millions of lines of software code and fault-tolerant distributed hardware. The increased complexity of telecommunication systems has made them very difficult to verify. With the advent of the personal computer revolution, a new element of complexity has been added to the design. Figure 1.1 illustrates the design complexity. Traditional voice-based services have emphasized the design of highly complex digital switches. While the design of digital switches is not a straightforward process, the complexity was managed by the manufacturer. The new users demand more broadband capability, and emphasis is now being placed on the design and integration of devices such as ATM. Although such devices are not individually as complex as the digital switch, a network of such devices is. This is an important consideration, since a large portion of the traditionally centralized protocol management functions is distributed across the network. For example, an ATM-based network's survivability and error-recovery processes are possible at a network level, whereas this is managed on an individual basis with digital switches.

How does all this affect the design of such systems? A systems designer cannot afford to create a new device without considering the impact on the network. This implies that the design process must incorporate good modeling and simulation facilities for testing the new ideas with a representation of the network elements.

A typical design flow for a large system is represented in Fig. 1.2. The figure highlights the high-level flow of information from design concept to implementation. Four phases of a product's early life cycle are also shown.

- The *specification phase* is where the requirements and behavior of a conceptual system are first formulated. The result of this phase is a document and/or model reflecting the desired characteristics of the system.

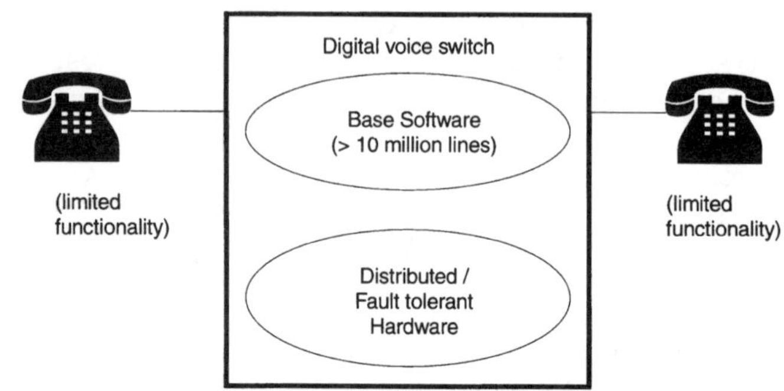

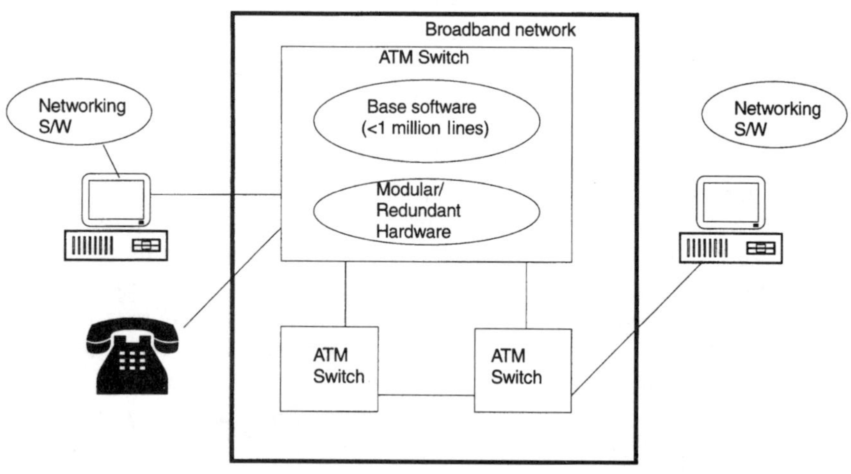

Figure 1.1 Complexity in voice versus broadband systems.

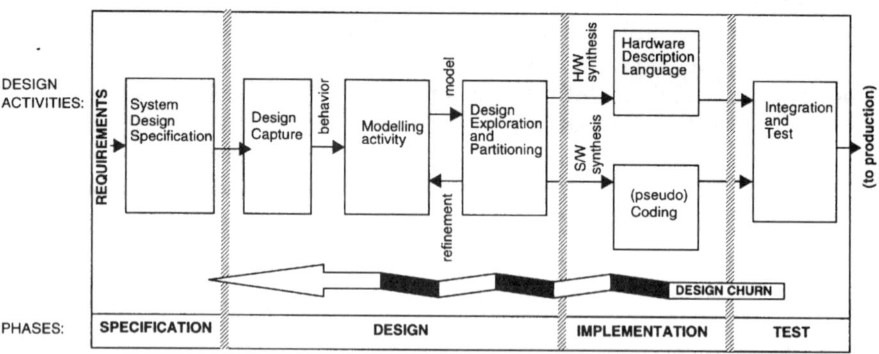

Figure 1.2 Typical design flow for complex systems.

- The *design phase* is where different solutions that meet the desired characteristics of the system are explored and evaluated. Typically this will involve the creation and simulation of different models of the system. The eventual goal is to produce a description of the design which complies with the specifications. Ideally, the description will be executable, so that it can be automatically used within the next phase.
- The *implementation phase* consists of all the activities that contribute to the physical manifestation of the design. The result is the first version of the product. This is generally an intermediate release of the final product.
- The *test phase* covers all activities that validate the product's behavior against the specification. The product is then ready to be released to the customer for acceptance testing.

The phases do not necessarily have well-defined boundaries—for example, specification and design may be considered as one phase within a given engineering group. It should also be noted that there are many micro-level activities which are not shown, such as unit testing, design reviews, and model iterations. A deliverable associated with one phase may be made available to other phases even before the completion of all activities within its phase (the infrastructure to support the design and management processes is referred to as concurrent engineering). Listed above each phase are some of the related design activities that contribute to the overall time required to complete that phase.

Initial system requirements in the specification phase generate design specifications that are input to the design phase. The design phase encompasses many activities related to modeling, design exploration, and synthesis. The modeling activity can be composed of model creation and analysis. A design exploration activity can use the model to simulate different scenarios or apply formal verification methods to obtain an acceptable representation of the system. The designer can also partition the design into the respective hardware and software components and, through the use of synthesis tools, generate a prototype (or product) in the implementation phase.

For hardware designers, description languages exist which can represent systems at various levels of abstraction. The proliferation of these languages has prompted the need for standardization (IEEE 1076 VHDL and Verilog) (VHDL 1987; Verilog 1991) to promote design migration and reuse. Current commercial design environments permit synthesis of hardware described in a subset of VHDL or Verilog, where the initial design is at a register-transfer level of abstraction. The software developers may utilize methodologies such as Shlaer/Mellor (Shlaer 1992) or Ward/Mellor (Ward 1985) to describe the software constructs in more detail. Each phase in the product life cycle can potentially flow back to the previous phase(s), indicating potential design revisions. This creates design churn and adds to the overall time to deliver a product.

The activities and tools used within a typical development process may differ in each organization, depending upon the organization's *maturity* level.

Organizational maturity

Organizations can be described as having attained a certain level of development *maturity*; this is the level of competence that the organization has reached as compared to other organizations. Several models exist in the industrial engineering field where levels of maturity are defined in terms of the tools, techniques, and processes that the organization has in place to produce a product. *Trillium* (Bell Canada 1995) is an example of a known practice for describing maturity levels within the telecommunications domain. This practice divides an organization into eight capability areas, each of which defines some aspect of the organization (see Table 1.1).

Each area is rated at a capability level, which is a five-point scale reflecting the maturity of the organization in that area. The scale rates organizations as being unstructured (level 1), project-oriented (level 2), process-oriented (level 3), integrated (level 4), or fully integrated (level 5). Most of the current development organizations are

TABLE 1.1 Trillium Capability Levels

Trillium capability area	Contains practices at level				
	1	2	3	4	5
Organizational process quality					
Human resource development and management					
Process					
Management					
Quality system					
Development practices					
Development environment					
Customer support					

SOURCE: Bell Canada, *Trillium Rel. 3.01, Model for Telecom Product Development and Support Process Capability,* June 1995.

either at or aspiring to achieve level 3 capabilities. Two of the capability areas, development practices and management, highlight some strong requirements for complexity management within the design process. In particular, a level 3 organization is expected to have prototyping processes in place. These are defined in Trillium as follows:

> Prototyping techniques are used to explore design options and validate design decisions prior to implementation.

Level 4 requirements stipulate the use of formal methods for development of critical components, computer-aided systems engineering (CASE) tools, and design metrics. Level 5 organization characteristics are automatic code generation, executable generation, and constructive development. The latter is defined as follows:

> Constructive development methods (i.e., methods in which concrete intermediate products are constructed from abstract intermediate products) are used for most development tasks.

It becomes evident that the maturity of a design organization is related to its ability to manage complexity. The design tools of advanced groups will include rapid prototyping, formal analysis, executable code generation, and constructive development methods.

Design automation

Intuitively, the time period required to create a product (from requirements to test) can be influenced by using integrated tools to shorten the various phases. The tools can either reduce the duration of activities (such as modeling) within phases or reduce the design churn time. Traditional design approaches have focused upon such tools. Integrated across several design activities, the tools can form a rapid system prototyping (RSP) environment to quickly produce a prototype system, which is a scaled-down version of the final product. The prototype is a model of a conceptual system to be fabricated and relies upon the environment to expeditiously produce sound systems.

Although automation is a way to shorten the design time, there are other reasons to apply it as early in the product life cycle as possible. The use of tool capabilities for initial specifications promises to reduce design ambiguities and errors early in the product life cycle. Such tools must be capable of design capture at increased levels of abstraction to support modeling at the *architectural* design level. The architectural level is the most abstract design level defined by Bell (1971). This representation is given in Table 1.2. At the architectural level of design abstraction, the behavior of system elements is viewed as communicating processes. The elements may be described as processors, memories, switches, or other high-level descriptors.

TABLE 1.2 Levels of Design Abstraction

Level	Behavior	Structure	Physical domain
System (architecture)	Communicating process	Processors, memory, switch	Cabinets, cables
Algorithm	I/O	Memory, ports, processors	Board, floorplan
Register-transfer level (RTL)	Register transfers	ALUs, regs, MUXes, buses	ICs, macro cells
Logic	Logic equations	Gates, flip-flops	Standard cell layout
Circuit	Network equations	Transistors, connections	Transistor layout

SOURCE: C. G. Bell and A. Newell, *Computer Structures: Readings and Examples,* McGraw-Hill, New York, 1971.

Bell's original definition of the design hierarchy labeled the *system* and the *architectural* level as equivalent. Since then, the term *system* has been overloaded to the point where it has no well-defined global meaning. For example, the term is utilized by individuals working at all the different levels of abstraction, from circuit designers to component integrators. Consequently the word is interpreted differently in the various communities. The term *architectural level,* on the other hand, has not suffered such a fate. It is used less frequently, and as a result its meaning is still relatively untarnished. In this book, the term *architectural level* explicitly refers to the design abstraction layer that is above the algorithmic level and below the conceptual representation of a design, as described by Table 1.2.

RSP environments utilize an internal modeling or specification language which describes a given system at the desired level of abstraction. The language must also be sufficiently versatile to accommodate the exploration of the design space and further refinements to the design. In the case of architectural modeling, this would also involve partitioning the design into its constituent hardware and software components. This activity may utilize simulation and formal methods as design tools. Finally a synthesis activity uses the refined design to produce the target product. The source and target model for synthesis reflects the level of synthesis being undertaken in the design. The four different design abstraction levels are commonly depicted as concentric circles radiating from the center of a three-dimensional graph (Gajski 1992), as shown in Fig. 1.3. The intersection of these circles with the three axes represents three possible views of design domains (behavioral, structural, and physical) at a given level of abstraction.

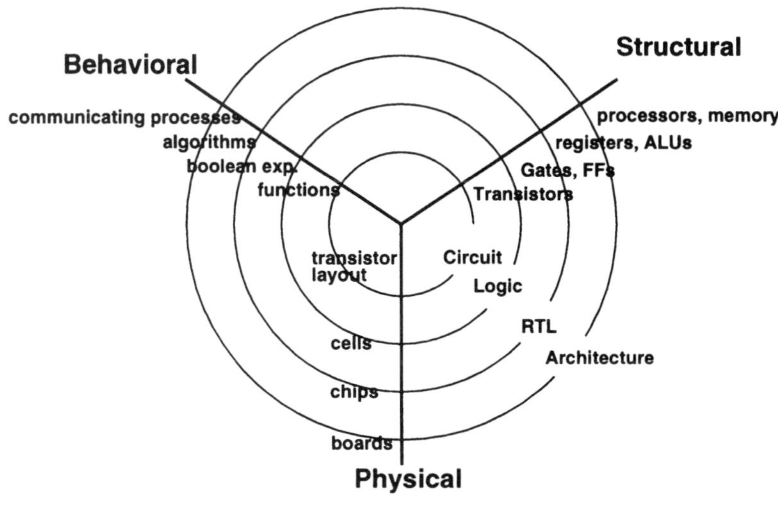

Figure 1.3 The Y diagram for synthesis (Gajski 1992).

As the circles move away from the center, the level of representation becomes more abstract. Synthesis can typically occur from a given behavioral representation to a structural one.

The Cost of Errors

Moving tools up the abstraction scale is attractive for several reasons. It is well accepted that design errors uncovered early in the life cycle are orders of magnitude less expensive to fix than those detected downstream (Boehm 1981). Figure 1.4 depicts the relative cost of fixing errors at different stages of development. As can be seen, the cost of fixing errors later on is orders of magnitude greater. Furthermore, as the complexity of a system increases, the cost becomes even greater.

Early error detection can be a significant benefit of architectural-level automation. Figure 1.5 further emphasizes the impact of design choices made in the early phases. The figure depicts product development in terms of a project within a corporation (Saultz 1992) and illustrates the typical resource and personnel effort requirements during a product life cycle. The horizontal axis depicts the different phases that the product may progress through, while the vertical axis indicates the percentage of the total cost of the project in terms of total effort and resources (equipment, personnel, money).

The diagram demonstrates that more than half of the total cost has already been committed after the specification phase, whereas only 15 percent of the cost is incurred. It should also be noted that resources

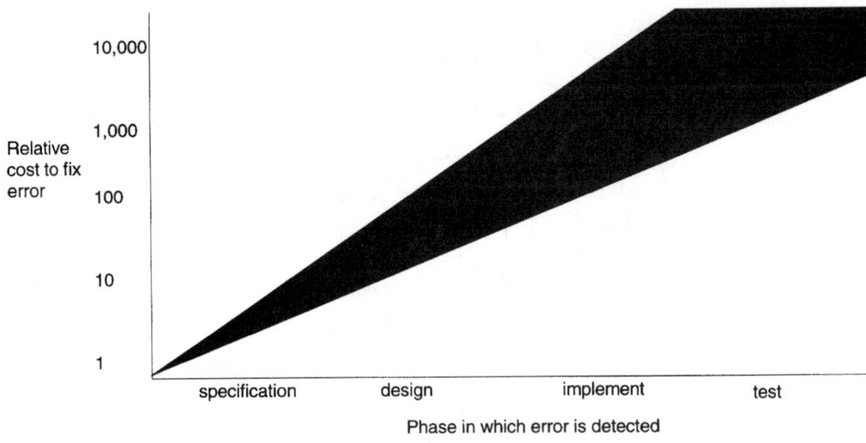

Figure 1.4 Relative cost of errors.

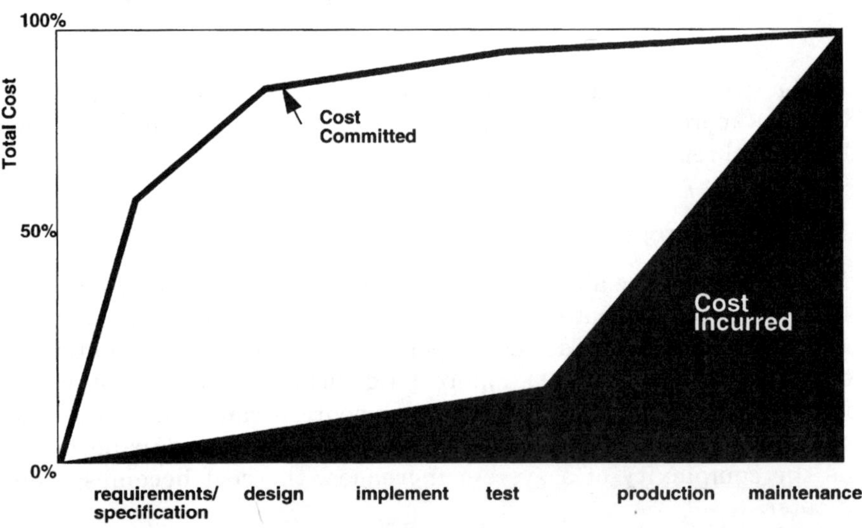

Figure 1.5 Typical project cost demands (*Saultz 1992*).

freed from previously completed phases may have been allocated to other projects. This makes them difficult to access for potential design reworks. Before the project reaches the production phase, nearly 90 percent of the total cost has already been committed. As a result, significant design errors detected downstream can negatively affect the resource allocation to the project, resulting in cost overruns and delays. Hence the resolution of design ambiguities at an early stage ensures that potential design reworks and churn are reduced, so that the product can be developed with the limited resources available. Hence, using design aids at an earlier point in the product's development can significantly curb development costs (Hayes 1988).

In summary, an RSP environment applied to architectural-level design of large systems can have a significant impact upon the design cycle in areas such as:

- *Shorter design cycle.* Generating a quick yet reasonable design from the very conception of the product can weed out poor design decisions. This reduces expensive design reworks later in the design stage. As a result, the time from system conception to integration is greatly reduced.
- *Verification of designs.* Current designs are verified later in the design stage. At that point, poor architectural decisions can be hidden in the complexity of the complete design details. By synthesis of a system design, one obtains a formal representation or model. Algorithms or rules can be applied at this level to test the correctness of the design against the design specifications. Ideally, synthesis would generate designs that are correct by construction. This would lead to fewer design errors later on in the design cycle.
- *Solidification of design requirements.* Vague or ambiguous design specifications can be identified and corrected during this early stage. Traditionally, designers have interpreted the specifications, but their interpretation may not necessarily have been what was intended (Srivas 1990; Moore 1990). The level of abstraction in the architectural level is closest to the conceptual model that a designer would work on, minimizing loss of information.
- *Procedures for formal design specification.* With the advent of system-level synthesis tools, formal requirements for design specifications can be defined. Formal specifications following a well-defined structure or methodology are currently lacking in the design community.
- *More design alternatives.* If design space exploration can be performed quickly, different design alternatives that might otherwise not have been feasible can also be explored. This helps guide designers in the conceptual creation or evaluation of their designs.

Environments

Design and prototyping environments have been developed mostly to address synthesis at register-transfer and lower levels. For example, IDEAS (Kumar 1989) is an environment that allows RTL synthesis of designs. CATHEDRAL (Lanneer 1991), ISPS (Barbacci 1981), and

HIS (Bergamaschi 1993) are examples of high-level synthesis tools. As the lower-level design tools have matured, research focused on architectural-level synthesis (transforming a system-level design specification into an algorithmic-level specification of the behavior) is finally starting to become feasible and is drawing more attention. Some success with reasonably complex systems has been obtained for systems dedicated to specific problem domains such as digital signal processing, where the design space is relatively homogeneous (Lanneer 1991).

With the increased level of abstraction, the design space for a "general-purpose" architectural-level RSP tool is heavily heterogeneous, making it infeasible to derive a model from most initial specifications (Ramming 1993). As a result, it is generally agreed that successful architectural-level RSPs will be domain-specific (exhibiting a reasonably homogeneous design space) or support a wide variety of specification paradigms. Figure 1.6 shows the difference that is implied between the approach taken by architectural-level modeling and a traditional one. In many instances, modeling and simulation for a given design is performed across many groups. However, the information and knowledge gained from these activities are likely to be retained only by the individual groups. The models are generally not portable or reusable between groups. For example, a model of the hardware is hardly ever seen by the software designers, so that potential design tradeoffs must be performed either before or after the individual design. In the architectural-level design, the system is designed at a high level of abstraction involving both software and hardware elements. Models can be reusable in the sense that past designs can be incorporated into new ones, permitting exploration of more design alternatives. Domain knowledge is also used as input to help direct the outcome of the design.

Figure 1.7 reflects the change in the role of simulation in the design of systems. In an organization consisting of software and hardware designers, more time is dedicated to modeling and simulation using an architectural approach. However, the anticipation is that there is less possibility of errors incurring downstream in the development process.

The demands upon the modeling capabilities of an RSP environment imply that the internal representation or design language employed by the environment can greatly affect the acceptance of the tool within a given domain of application. At the architectural level of abstraction, design partitioning and codesign are major activities involving both software and hardware designers. Therefore, a design language is needed that will accommodate both disciplines transparently.

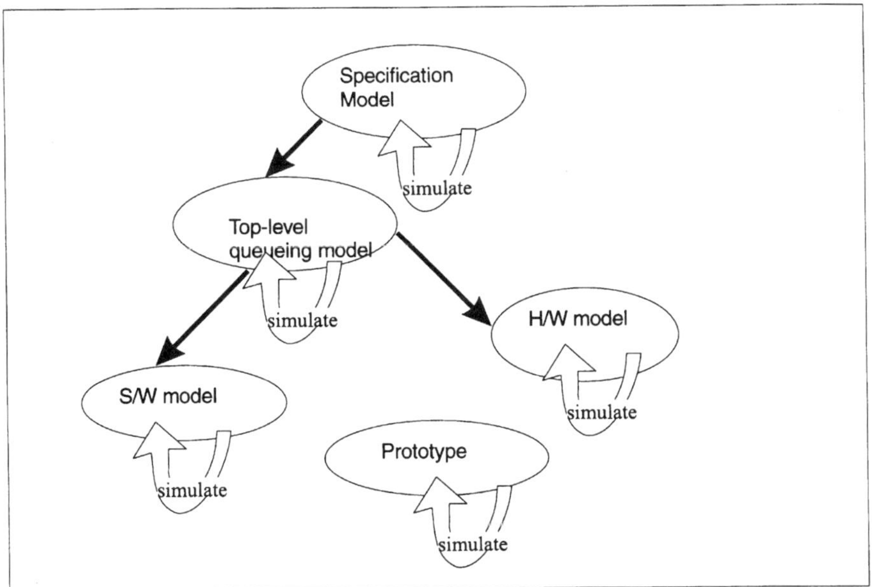

Typical simulation/models from large design groups

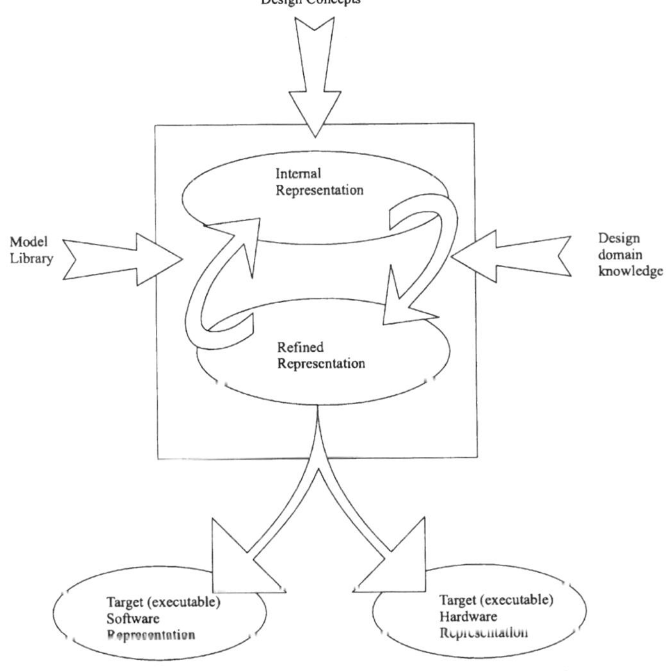

Architectural level environment outputs

Figure 1.6 Traditional versus architectural model simulation activities.

Typical design environment

Specifications	High-level design	Implementation	Simulation	Test/ Error Fixes

Architecture level model based environment

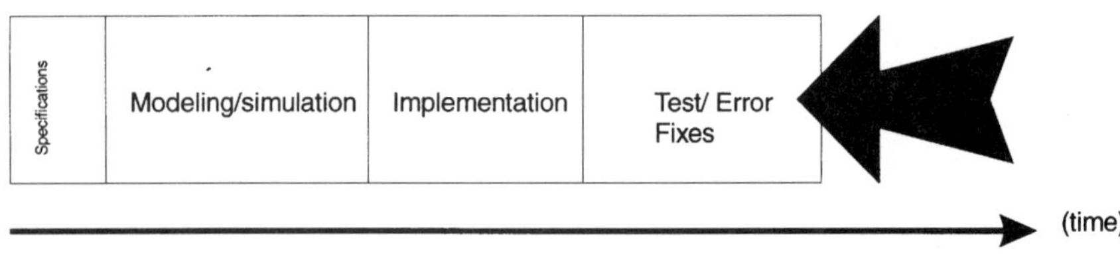

Figure 1.7 Timeline for simulation activities in design.

In the hardware community, design languages have evolved from logic-gate-oriented formalisms to languages capable of modeling up to the algorithmic abstraction level. However, they rarely support architectural-level representation [although attempts are being made to extend VHDL to this level (Jerraya 1991)]. Most hardware description languages are simulation-oriented and interleave simulation and modeling semantics, making them difficult for use in synthesis systems. An example is VHDL, where only subsets of the language are applicable for synthesis by commercial tools.

Within the software design area, the trend has been toward a parallel and concurrent programming view. This has made the use of specification paradigms that support communicating processes highly amiable. A similar trend can also be observed within the hardware design community, with the desire to find better ways to specify the existing parallelism inherent in hardware. Hence, at the architectural level, where software and hardware design concerns are first addressed, a formalism based upon communicating processes appears natural to designers from both disciplines (Koomen 1991). For example, the Specification and Description Language (SDL) (CCITT 1988) is based upon communicating processes and is in use in telecommunication software development groups (Klick 1991; Jacobson 1992). Although descriptive, the language is undergoing revisions to adopt object-oriented views and a timing model. Another language is Statecharts (Harel 1987), which allows description of systems in terms of hierarchical communicating finite-state machines expressed in a graphical notation.

Outline of Book

The book is structured in nine chapters. The first half of the book deals with the necessary ingredients of a modeling and design environment for complex systems.

- Chapter 2 presents an overview of basic design issues. Artistic creativity versus scientific control over the design process is elaborated. The different models for the development of products are reviewed. Design basics which are crucial to the legibility of the book are introduced. An overview of different representation schemes, such as Petri-nets, SDL, and Prolog, is also included. The chapter closes with a look at fundamental concepts of simulation and domain-directed design.
- Chapter 3 introduces the ideas related to architectural-level design. This approach will lay the foundation for developing the modeling framework for complex systems. An example of a digital telephone switch is introduced.
- Chapter 4 introduces a language which can be used as an intermediate representation form for architectural modeling of systems. The telephone switch is described within the language's notation.
- Chapter 5 examines the issues related to the execution and simulation of the models. The chapter identifies means for analyzing models and the basic data structures required by the simulator.
- Chapter 6 is dedicated to design synthesis issues. In particular, translation of models from the intermediate form to a hardware description language (VHDL) is presented.
- Chapter 7 presents implementation issues related to the creation of an architectural design environment. The intricacies related to constructing a prototype environment (Design Analysis and Synthesis Environment) are presented to illustrate the concepts.
- Chapter 8 presents a detailed modeling and design exercise for an ATM switch and a broadcast protocol model on a network of ATM devices.
- Chapter 9 examines some of the outstanding advanced concepts, such as model library issues and design space exploration.
- The appendices provide complete listings of code related to the examples described in the text.

Chapter

2

The Art of Design

This chapter examines some of the fundamental elements that play a part in the design process. The goal of the chapter is to sensitize the reader to these by providing a snapshot of current practices and, consequently, present a premise for an architectural-level modeling of complex systems. The chapter proceeds with a general comparison of creativity versus process attributes in the design cycle. An overview of the different models used to depict the design process is presented. The chapter then provides an introduction to some of the tools and concepts used in the design of systems. These are so abundant in the literature that the applicable formalisms and techniques available to a designer cannot all be fairly covered in one chapter. Consequently, as a compromise, the methods that will be discussed in this chapter were chosen because they provide a basis for the formalism that will be utilized in the later chapters. The reader familiar with these basic design concepts may wish to proceed directly to the next chapter.

The design of telecommunications systems, like that of other complex systems, is a process that requires a significant amount of discipline and creativity. In most cases creating a design similar to existing ones may not be sufficient; it may be necessary to actually improve upon existing designs. The establishment of a design process is one of humankind's earliest achievements. The basic principles used in realizing centuries-old ideas are still valid today. In its basic form, to design something implies to devise, create, fashion, or construct it according to a plan or specification. Essentially the process is initiated with a driving force: an idea or specification. This is transformed, through the designer's creativity, from a concept to a tangible form. For larger and more complex systems, a validation step is typically used to test the concept before it is fully constructed.

Art and Science

The design activity is sometimes referred to as being *art* rather than *science*. Purist engineers and scientists may scoff at the notion that design can be termed an "art" rather than a well-defined process. After all, products have been designed in relatively the same manner for years in a reasonably controlled fashion. The advocates for the acceptance of "art" would argue that a great deal of creativity is involved in transforming the conceptual idea into a realistic design—somewhat akin to a sculptor's chiseling out a concept from a slab of marble. The designer may weigh different choices, modifying previous decisions as the design is being molded, and finally settling upon a solution which fits within the confines of the original specifications.

In fact, both claims are valid. There is an element of creativity involved in an otherwise scientific process. In fact, a creative designer can often provide an added element in a design, making it unique and more competitive. However, while companies like to employ creative designers, they also prefer well-disciplined design reviews and test procedures. A continuous balancing act is carried out in design organizations to ensure the right combination of artistic and scientific resources. Hence the main difference, apart from the obvious, between the creation of a sculpted masterpiece and a well-designed digital system can be attributed to the level of artistic control. A sculptor has virtually complete artistic control over the end product. One consequence of this is that a sculptor is limited only by his or her own creativity and the tools of the trade. Another consequence is that it is extremely difficult for another sculptor to create the identical sculpture with the same tools.

In the case of a typical digital systems design, most of the artistic creativity is employed during the early stages of design, when the initial architecture is being contemplated. The design is then transferred into some machine-executable representation, after which it follows a relatively well defined process until implementation. Hence it is repeatable, independent of who is involved in the process.

The basic goal of the design process is presented in Fig. 2.1, and it is important to keep this perspective when dealing with some of the more involved issues discussed in later chapters. Essentially, the horizontal axis in the figure depicts the traditional design activities or stages that are followed in realizing a design. The terms used in the figure are those typically used in large-scale telecommunications designs, but those used in other domains can be substituted. The vertical axis shows the increasing levels of abstraction, eventually cumulating in an intangible level which will be termed *concept*. This limit essentially refers to the first describable notion of a design. Once a

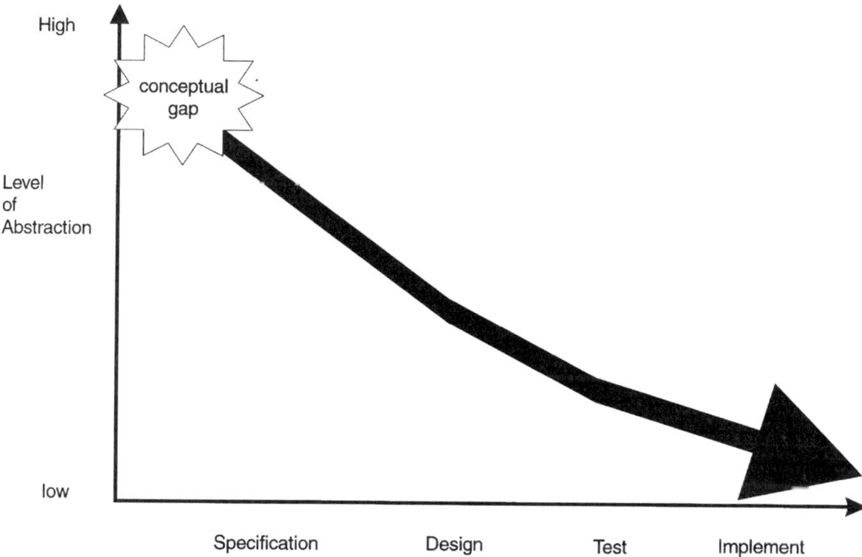

Figure 2.1 Objective of a design process.

concept is generated, it can be interpreted by the individual as design requirements and then design specifications. Subsequently, the specification can be executed, analyzed, and revised during the execution phase. Finally, it can be tested and then implemented as a physical product. The curve in the diagram essentially depicts the main driving force of the whole process: the derivation of a physical realization from a conceptual one.

The diagram is admittedly simplified. There are points in the process that may require iteration, but the overall flow or evolution of activities over time is the same. What is often forgotten is that whatever errors are injected early in the design process are propagated and compounded to later stages. In fact, the section in the figure marked "conceptual gap" is a serious issue that both hinders and fuels complex system design. This gap implies a need for significant human intervention in the normal design flow, resulting in creative or poorly designed products, depending upon the expertise or mastery of the human designer. Unfortunately, there is no way at this time to extract the conceptual form of a design from the thoughts of an individual and objectively transfer it into a machine-understandable form. Consequently, the translation from concept or requirements to specifications is prone to errors.

The conceptual gap is not unique to digital systems design. The sculptor may transfer the concept (the idea to be sculpted) into a sketch of the concept (the specifications) and then start sculpting. If the sketch does not convey the meaning of the initial concept, then the resulting product may not be what the artist originally envisioned.

Traditionally, the conceptual gap has been unconsciously accepted as a necessary evil. Design aids that assist the designer to make system-level choices by rapidly evaluating different design options are scarce. Such tools would permit designers to use initial design requirements or specifications to engage in a design space exploration. Unfortunately, as was indicated in the previous chapter, traditional development processes favor test and simulation after design rather than before. At this point, it is worthwhile to review the major development process models to understand where the design activities fit in the greater scheme of things.

The Product Development Process

There are several models for describing the product development process. Some of the better-known ones are reviewed below. However, for a more complete exposure to the area, the reader is urged to further examine the references.

Waterfall model

A typical waterfall model is depicted in Fig. 2.2. Slight variations in the model are possible, particularly in the naming convention and number of phases. This model is one of the earlier attempts to describe the logical progression of activities that take place in the development of a product. The model begins with the requirements specification phase. This activity is typically initiated directly or indirectly by the customer for the end product. The result of the activity is a documented description of the basic specifications of the system. The next phase is the analysis of the specifications to produce a more technical description of the system. This is then utilized in the design phase, where different components of the system are conceived. The system components are implemented, and subsequently individually tested. Upon successful test completion, the system components are integrated and (usually) undergo a system-level test. The product is then released and undergoes a maintenance phase.

The waterfall model is a rough approximation of reality. In particular, it was soon evident that in most cases, feedback arrows connecting each stage to previous stages would be required in order to reflect design iterations. This model also failed to describe some of the complicated dynamics in the individual phases. For example, the design

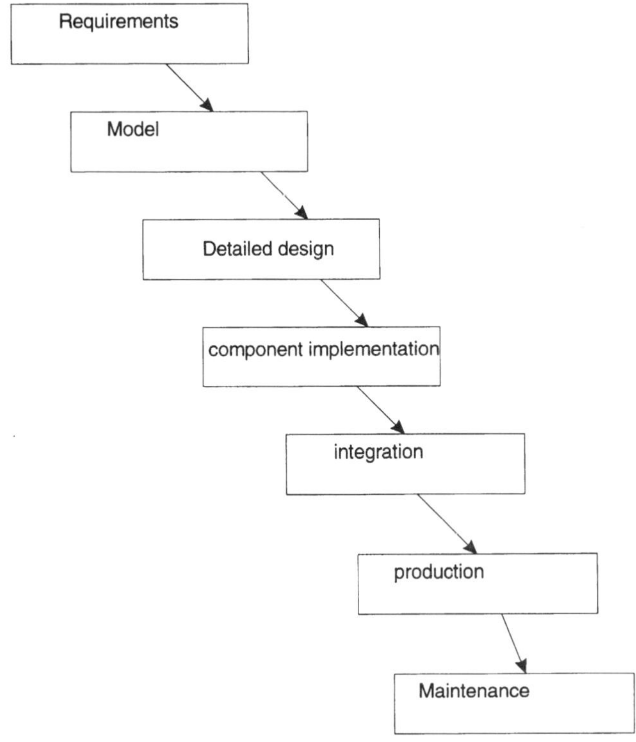

Figure 2.2 Waterfall model.

and maintenance phases can each employ an iterative internal process similar to the waterfall model itself!

The spiral model

To capture the interactions missing in the waterfall model, the spiral model, shown in Fig. 2.3, was proposed (Boehm 1986). The activities depicted by this model progress outward in a spiral fashion. The activities along the spiral progress through an iterative series of specifications, intermediate designs (prototypes), and analysis, eventually produce a final design. Each rotation of the spiral corresponds to increasing costs and time. However, since the number of rotations is dependent upon the number of intermediate prototypes generated, which is generally not known beforehand, it can be difficult to plan the time and cost of the project using this model.

Even with its shortcomings, the spiral model provides a relatively accurate portrayal of the product development cycle.

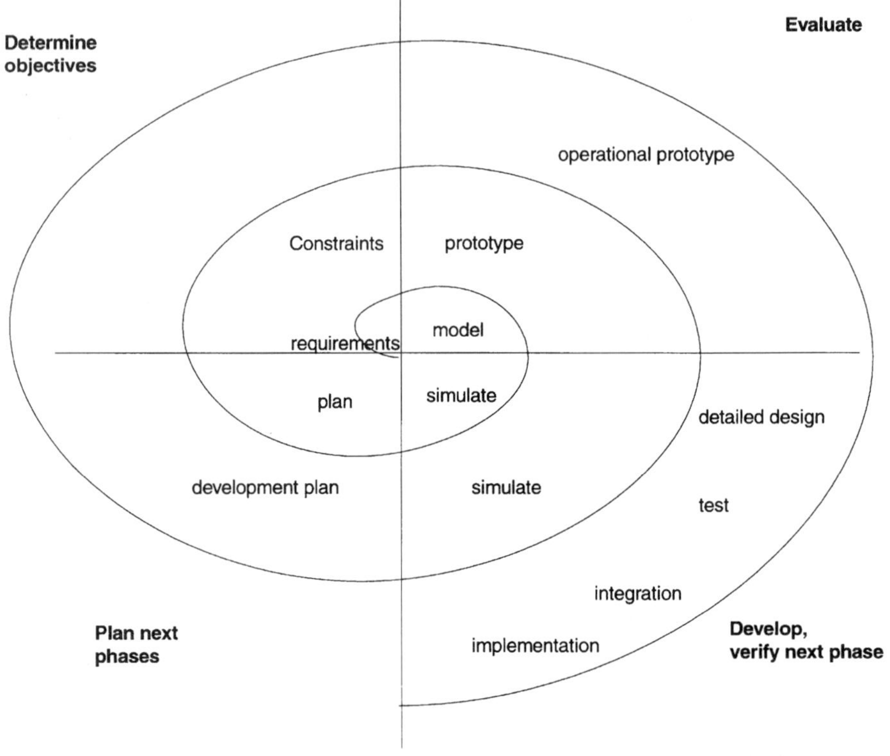

Figure 2.3 Spiral model.

Rapid system prototyping

The evolution of design automation has made it feasible to generate fast intermediate designs that emulate different portions of the overall system before committing to the final design. This approach permits testing different design options against the specifications. In most cases it involves the creation of a mock-up or simulated design with which the final user can experiment.

A typical rapid system prototyping (RSP) process is depicted in Fig. 2.4. The process uses design specifications to generate a prototype, which is then scrutinized by the user. When the designers and users are satisfied with the performance and features of the prototype, the design is generated and implemented. Conceptually this approach is appealing, since an initial investment early in the process can help reduce design shortcomings later on. Hence the approach can ensure a good match between customer expectations and the final design.

There are also important considerations to be aware of when using the RSP approach:

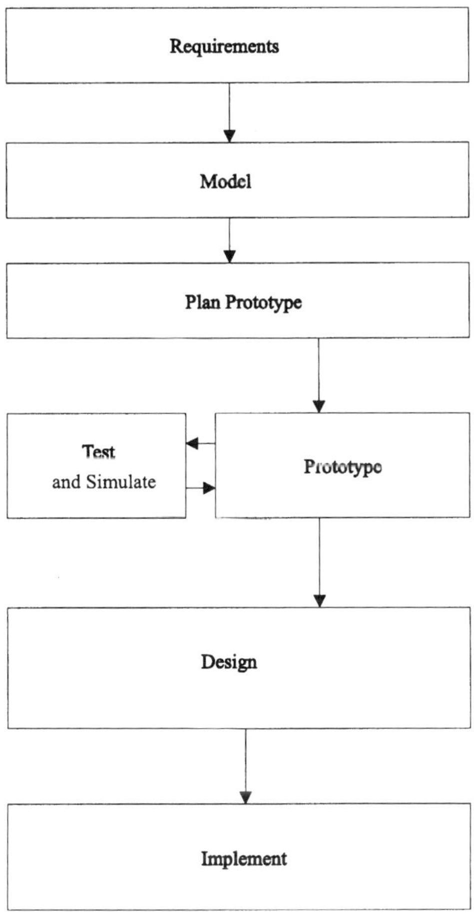

Figure 2.4 Prototype model.

- The traceability of the final design to the prototype is important. This implies that design changes are reflected back to and tested in the prototype. A common shortcoming is to use the prototype only as a means for obtaining customer acceptance for a given design and then discarding it. If the design undergoes several iterations after the prototyping stage, the final design may not meet some of the requirements established in the prototype. This defeats the purpose of constructing the prototype in the first place.
- Generally prototypes concentrate on features and not on the engineering performance of a system. Hence, even though features may *feel* right in a prototype, performance concerns (such as response times) may still arise after the design has been completed. A possi-

ble solution is to incorporate performance models and attributes into the prototype, so that the various performance parameters are also reflected downstream in the design phase.

Cooperative design

Design methodologies providing an integrated view of both software and hardware processes can be loosely termed cooperative design or codesign. The idea is that all processes are interlocked at key points in design so that both software and hardware issues are addressed together. Figure 2.5 shows the basic design activities undertaken.

The figure shows various points at which hardware and software design teams would need to synchronize their activities. However, an interesting element is the extra effort expended in modeling the system as a combination of software and hardware elements in the system-level model stage. Activities at this stage typically involve simulation and analysis of a high-level model. Hence the feasibility of different design tradeoffs and alternatives can be ascertained. Once the model is accepted by the design teams, a partition is performed on the design, breaking it into its respective software and hardware components. The teams typically progress through their respective development phases until an intermediate implementation is produced. At this stage, a system integration is undertaken to produce a beta version of the product (or a preproduction version), which then proceeds through the normal test, production, and maintenance stages of the product's life cycle.

In reviewing the basic steps in the cooperative design approach, it becomes apparent that the emphasis is on bringing the software and hardware designers together early in the process. This ensures a good global perspective on the entire design, with the necessary discussion on design tradeoffs. An additional time and resource burden is placed on the design teams because of the extra work involved in creating a system-level model and partitioning the design, but the long-term benefits include fewer surprises during the integration and test phases.

Design Basics

Given the basic role of design in the product development process, it is important to understand the basic concepts that are applicable to design activities. These pertain to the definition and construction of abstract representations of the behavior and structure of the system. This section will provide a generic description of the concepts.

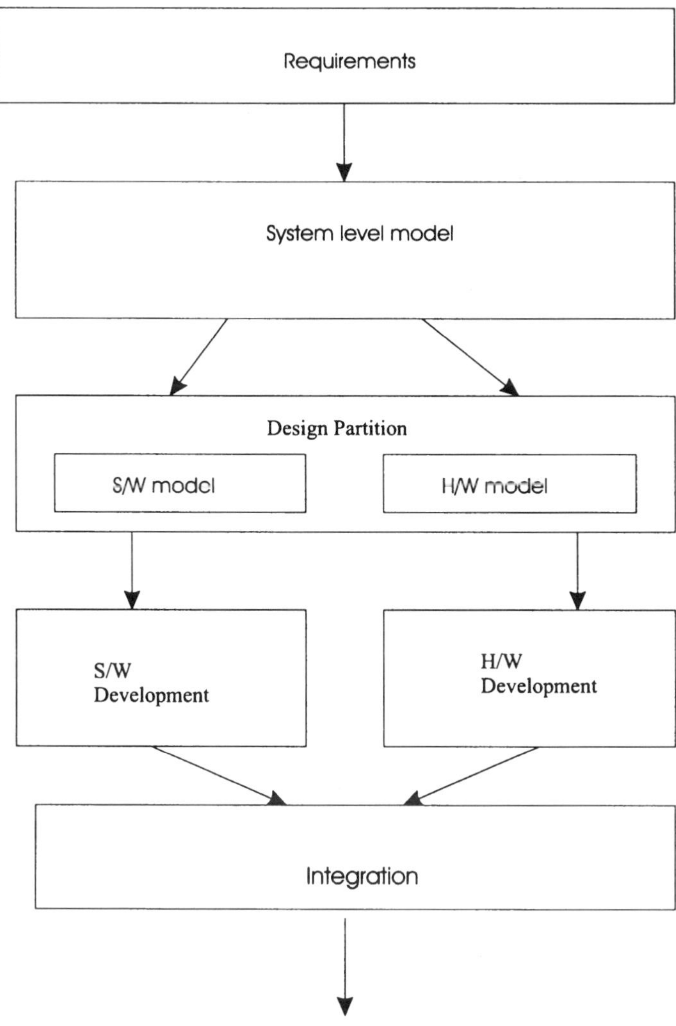

Figure 2.5 Codesign process flow.

The model abstraction

A fundamental notion in design is the use and construction of *models*. Within the scope of this book, a model will be defined as follows:

> DEFINITION: A *model* is an abstraction describing known or inferred properties of a system to be designed.

Alternatively, models are representations of the *behavior* and *structure* of a system to be designed. It is generally agreed that modeling is

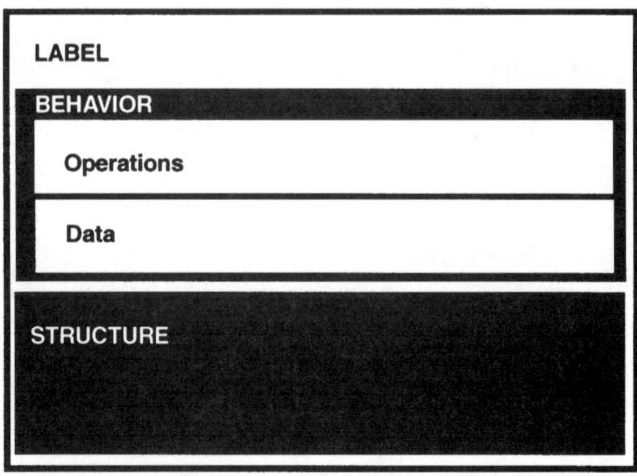

Figure 2.6 Generic model abstraction.

the most basic activity of a designer, in which real or imaginary worlds are expressed in the symbol space of an underlying language. Modeling is even more abstract than coding in that coding is required only if a translation from the specification language in which the model is expressed to another language understood by a target machine is needed.

A model represents a segment of the domain being studied. It replicates the behavior of its real counterpart in symbol space by generating trajectories consistent with those of what it represents. A model is the most basic construct available to a designer, analogous to an object in the object-oriented paradigm. Therefore, support for defining a model in any design environment is a fundamental task.

A generic description of a model is given in Fig. 2.6. The model is defined to possess (at a minimum) a behavior and a structure. A model's behavior can be described using a set of operations on data. A data structure, often referred to as a state or representation, may typically take the form of a stack, list, or queue. Operations manipulate the data and need to be consistent with the structure.

The structure defines the interfaces which permit models to interact with their environments, including other models, in a predefined manner. Typically, communication is established by message-passing mechanisms.

Example. To illustrate the model concept, a simple example will be presented. Let us assume that we would like to define a model of an *adder-subtracter* unit. An informal functional description of the unit is as follows:

> The adder-subtracter unit accepts two inputs (*input_1* and *input_2*) and outputs the sum (at *output_1*) and the difference (at *output_2*) of the two inputs.

In a pseudocode form, the model could be defined as follows:

```
The structure of the model is
/** inputs: **/
input_1, input_2: integer: variable;
/** outputs: **/
output_1, output_2: integer: variable;
```

The behavior can then be defined as

```
output_1 = input_1 + input_2;
output_2 = input_1−input_2;
```

The above operations assume that the inputs have been previously instantiated with particular data values. Hence, if

```
input_1 = = 4;
input_2 = = 2;
```

then obviously,

```
output_1 = 6 and output_2 = 2
```

Composition

The model abstraction by itself is of little value. Models need to communicate with one another to exchange information or provoke a state change. Consequently, the composition of models rather than the creation of isolated models becomes more significant to the designer. With communication, complexity issues arise as the number of communicating models grows. Hence a structured hierarchical approach to model composition is amiable for large systems.

The model construct, viewed as a class in the object-oriented sense, is a natural foundation for developing representations of complex systems. The class can be reused in other models, inheriting all or parts of the attributes of its parents. For example, a model can be a superclass of a submodel by inheriting its properties. The benefits are numerous. Some of the primary ones are:

- Model libraries can be constructed and component models can be used and tailored to interconnect with one another in a multitude of ways. Hence interconnections can be easily changed and manipulated without affecting the underlying behavior of the component models.

- Reuse of models is encouraged through the support of a model library mechanism, as mentioned above.
- Composition of models enables the construction of a hierarchy of models. Composition supports the concept of hierarchy, reducing the complexity of managing and constructing large models.
- An increased degree of testability is implied. Models can be tested at component levels, and verification of their composite models becomes easier.

Building up models from simpler ones or decomposing a complex model into its submodels is extremely desirable in a modeling environment. Models can be constructed from previously defined submodels or from a composite of several submodels. The details that permit this level of construction are termed the *interface* specification of the model. A composite model can thus be viewed as one for which interface information on its submodels is maintained. Subsequently, the aggregate behavior of the submodels will reflect the behavior of the composite model. These can be regarded as an extension to the previously defined *model.*

> DEFINITION: A model composed of several interconnected models is termed a *composite model.*

The general structure of a composite model is shown in Fig. 2.7. The major distinction of composite models is the existence of an interconnection scheme to connect the interfaces of the component models. Composite models may have operations for pre- and postprocessing. These may be utilized to interconnect or configure the submodels associated with the composite models in a predefined manner. In many instances, the submodels must also be placed in a known state of operation.

With the composition of submodels, additional issues arise; these will be mentioned here briefly, but treated in more detail in later chapters. A problem of synchronization may arise when a simulator is scheduling messages generated from submodels. This appears as a conflict between two or more component models which are simultaneously scheduled for an internal activity. This problem can be resolved using a tie-breaking mechanism. Such issues are generally specific to the particular implementation of the model. For example, a distributed simulation environment may prefer optimistic or pessimistic scheduling techniques to maintain synchronization, whereas other environments may enforce an internal delta simulator clock delay before scheduling events, so that all events are eventually processed. The details of particular resolutions should not concern the modeler,

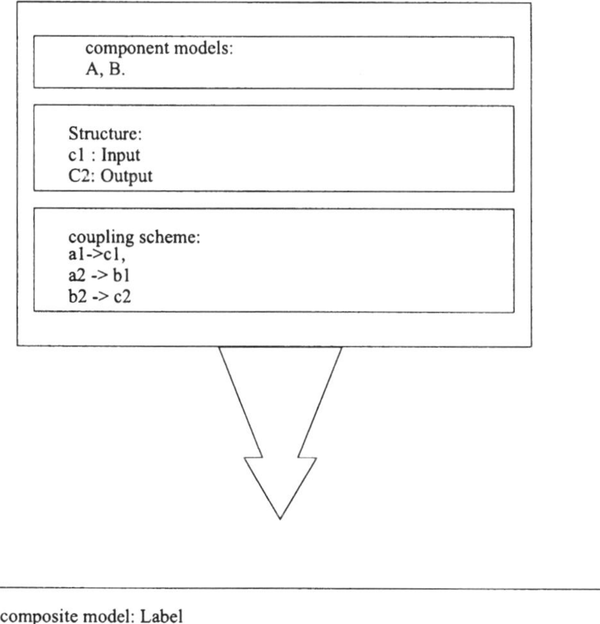

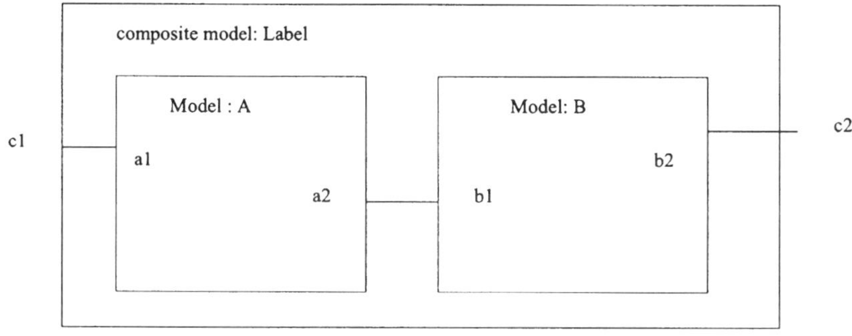

Figure 2.7 Composite model elements.

but the semantics should be clear so that the model can be structured to behave as intended.

Representation

The representation activity describes an implementation of the model abstractions within some formalism or syntactical boundaries. The language selected to implement the model will impose limitations, specific to the language's peculiarities, on the ease of representation. The representation can be distinct from the underlying implementation language, allowing the modeler to be mostly oblivious to the underlying language syntax. Three main representation formalisms are introduced in the following subsections that will be applicable in the later chapters.

Petri-nets

Petri-nets have been used in many areas to analyze hardware and software systems. The formalism provides for a relatively powerful way to define asynchronous concurrent communication. Since the original introduction of Petri-nets (Petri 1962), several extensions that increase their modeling power or expressiveness have been proposed (Jensen 1981; Jensen 1989; Murata 1989; Reisig 1982). Predicate/Transition nets are one of the extensions found in the literature. The following subsection will introduce the notion of a Petri-net and examine the Predicate/Transition net extension. The latter proves to be a useful paradigm for observing formal properties of the systems that will be discussed in the later chapters. However, other variants of Petri-nets could also be used.

Overview. An ordinary Petri-net graph (PNG) is defined as a triple PNG = $(P, T; F)$, where P is a set of places, T is a set of transitions, and F is the flow relation (elements of F represent directed arcs between places and transitions) such that $F \in (P \times T) \cup (T \times P)$, $P \cap T = 0$, and $P \cup T \neq 0$.

Markers called tokens move from place to place through the "firing" of a transition. The firing rule for a transition t in a Petri-net is satisfied when all input places contain at least one token. The execution of the firing rule results in the removal of a token from each input place connected to t, and the placing of a token in each output place connected to t. A Petri-net is said to be *bounded* if no sequence of firings can take the net into a configuration (marking) in which the number of tokens at any place exceeds a predetermined value. A special case is when the value is 1; then the net is deemed to be *safe*. Graphically, places are represented as circles and transitions as thick lines or rectangles. An example of a Petri-net is given in Fig. 2.8.

The figure consists of two places, which represent place holders for the inputs to the adder-subtracter unit. The number of tokens in the places corresponds to the values to be computed from the inputs. In the example, *input_1* is initially set to 2, and *input_2* is 1. After the net reaches a final state where no more firings are possible, the summation result is given by the number of tokens in place *output_1*. The difference of the two inputs is obtained from the number of tokens in place *output_2*. If there are no tokens in this place, then the number of tokens in place *P1* is the correct value, indicating a negative result.

The number of tokens in each place at any given time constitutes the state of the system and is referred to as the *marking* of the net.

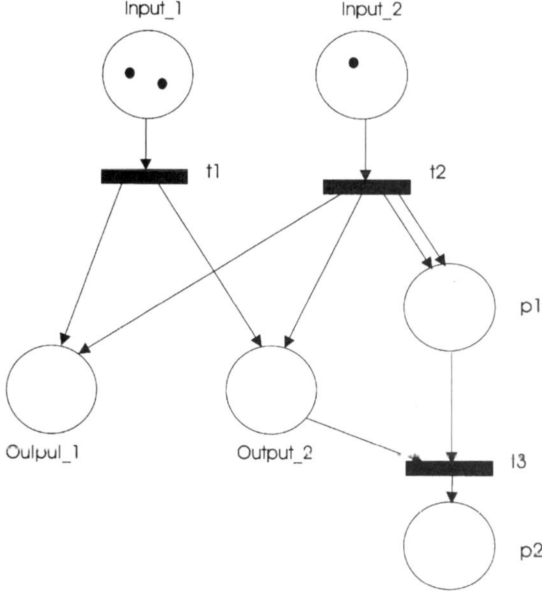

Figure 2.8 Petri-net example of adder-subtracter.

The initial marking of the previous example is defined as (2, 1, 0, 0, 0, 0), where the values represent tokens in places *Input_1, Input_2, Output_1, Output_2, P1,* and *P2,* respectively. By calculating all possible firing sequences, a *reachability* graph can be constructed, as shown in Fig. 2.9. The different markings have been labeled *m1* to *m11,* where *m1* is the initial marking. The graph indicates that 11 unique states of the system can be traversable, with the final state being *m11.* The reachability graph can be used to perform different analyses of the net. Terminal states can be detected (state *m11*) where no other firing is possible. The possibility of reaching a given state can also be determined by observing a path from the initial marking to the desired state. Also, bounds on places can be calculated or observed. Hence the graph can be used to verify if the number of tokens at particular places remains within certain bounds.

It should be noted that if a different number of tokens are used in the input places, the number of states or markings in the reachability graph will change. A common problem in Petri-nets is the state space explosion, where the number of possible states becomes too large to manage.

Another approach to net analysis is the use of matrix techniques. Basically, one can define a firing matrix *M,* where the number of

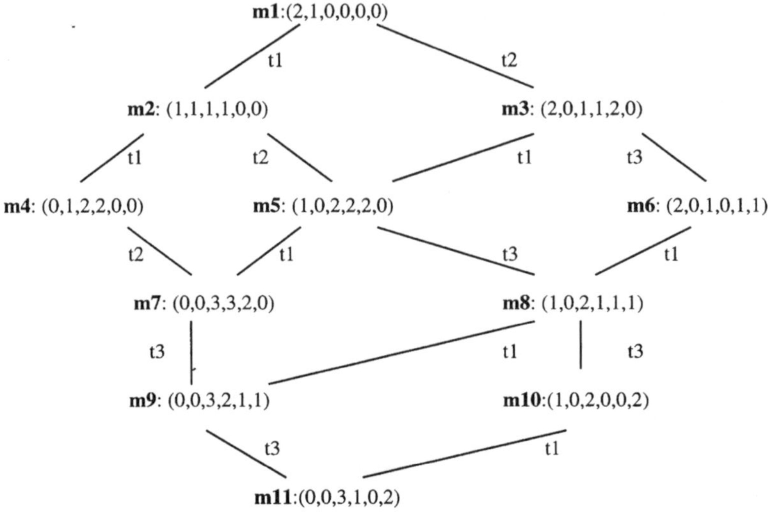

Marking of Places: (Input1,Input2,Output1,Output2,P1,P2)

Figure 2.9 Reachability tree for specific case.

columns is determined by the number of places, and the number of rows by the number of transitions. The entries define the number of tokens to remove (negative entries) or add (positive entries) to each place upon firing of a given transition. A vector *f* can be defined as being composed of zeros and ones. Nonzero entries indicate the transition(s) that are enabled. Alternatively, if a firing sequence is to be observed, a vector *fs* composed of the addition of all individual firings can be defined. Consequently, if *mr* is assumed to be a reachable marking, and *mc* is the current marking of a net, then the equation $mr = mc + fs.M$ must have a solution. The firing matrix for the example is

$$\begin{bmatrix} - & 0 & 1 & 1 & 0 & 0 \\ 0 & - & 1 & 1 & 2 & 0 \\ 0 & 0 & 0 & - & - & 1 \end{bmatrix} = M$$

where the column entries are *Input_1, Input_2, Output_1, Output_2, p1, p2* and the row entries are *t1, t2,* and *t3,* respectively. For example, given the initial marking $mc = [2, 1, 0, 0, 0, 0]$ and the firing of transition *t1* (i.e., $f = [1, 0, 0]$), then the next marking *mn* is

$$mn = [2\ 1\ 0\ 0\ 0\ 0] + [1\ 0\ 0] \bullet \begin{bmatrix} -1 & 0 & 1 & 1 & 0 & 0 \\ 0 & -1 & 1 & 1 & 2 & 0 \\ 0 & 0 & 0 & -1 & -1 & 1 \end{bmatrix}$$

$$= [2\ 1\ 0\ 0\ 0\ 0] + [-1\ 0\ 1\ 1\ 0\ 0]$$

$$= [1\ 1\ 1\ 1\ 0\ 0]$$

Using matrix manipulations, reachable markings can be computed. Once a marking is deemed to be undesirable, such as a failure mode in a system, then the conditions (the firing sequences) leading to the marking can be determined.

Predicate nets. An extension to the Petri-net formalism is the predicate (transition) net (PrTN). The following is an overview of the description of a PrTN based on the original definition in Genrich (1981); a complete formal description of the work can be found in the reference.

> DEFINITION: A PrTN is composed of an underlying PNG, a set of annotations A (applied to the PNG), and a representative marking M. Hence, PrTN = (PNG, A, M).

A is defined to represent four types of annotation, $A = (A_N, A_P, A_T, A_F)$:

- A_N is termed the support structure of the PrTN. It is composed of a finite set of constants U, set of variables V (ranging over U), functions fi in U, and relations Ri in U. It annotates the whole net rather than specific elements, describing static aspects of the net.
- A_P is a bijection between the set of places P and a set of variable predicates.
- A_T is a mapping of the set of transitions T, into the set of formulae (called transition selectors), employing only static predicates and operators.
- A_F is an arc labeling function which associates formal sums of tuples of variables and constants to the arcs. Hence the mapping is defined as $A_F \rightarrow W \cup W^2 \ldots \cup W^n$, where $W = U \cup V$ and W^k denotes the set of all k-tuples in W.

The marking M represents the displacement of tokens within the net. PrTN tokens constitute separate individuals, whereas in ordinary Petri nets tokens are indistinguishable from one another; only their count at places is of interest. Hence token sets are formed within PrTNs, attached to places and transitions of the net. If $x \in P \cup T$ and the arity of x is k, then a token set of x can be defined as $C(x) = U^k$. A given token $c = (u_1,..., u_k)$ in a place $p \in P$ denotes the fact that a predicate (of arity k) corresponding to the place is true for a given instantiation of the tuple of arguments contained in the token. Similarly, for a token c in a transition $t \in T$, the implication is that the variables within the annotations specified for the transition and its incident arcs are replaced by the corresponding constants appearing in the token.

The adder-subtracter unit is described as a PrTN in Fig. 2.10. Two places are used to hold the input values. The values are retained as an attribute of the tokens. The values are assigned to variables A and B as a precondition to firing the transition. The postconditions of the transition perform the addition and subtraction operations on the two values, and the results are deposited in the places *output_1* and *output_2,* respectively. It can be seen that the legibility of the graph compared to that of an ordinary Petri-net has greatly improved.

SDL

The Specification and Description Language (SDL) is a CCITT recommendation for specification of telecommunication applications. The language is based upon extended finite-state machines. An SDL specification is composed of processes, whose basic elements are states and transitions. A state represents a situation in which a process is in a passive mode, awaiting stimulus from an input. The input is in the form of a signal generated by another process. Once an input is received, the process executes a transition. A transition is a series of

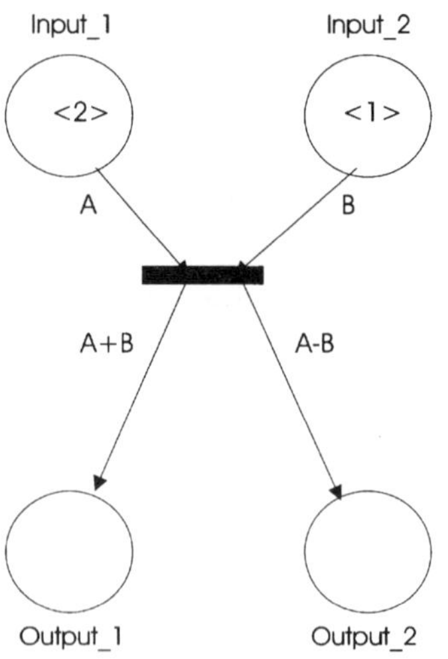

Figure 2.10 Predicate net of adder-subtracter.

actions that occurs when a process moves from one state to another. These actions can consist of signal generation (an output node), a set of operations (a task node), and decision nodes. Signals arriving at a process are implicitly queued in a first-in–first-out mode.

The adder-subtracter unit example is represented using SDL in Fig. 2.11. The process defined by the SDL model is assumed to reside in a state termed *start*. Upon receiving an input signal *input_1* or *input_2,* it makes a transition to the next state, *got_A* or *got_B,* respectively. Values associated with the inputs are retained by the process within the variables *A* and *B*. The process then waits for the next input. Once it obtains a second input, it proceeds to a task node to calculate the sum and difference. The results are then conveyed through signal output nodes as *output_1(sum)* and *output_2(diff)* signals. The process then reverts to the *start* state.

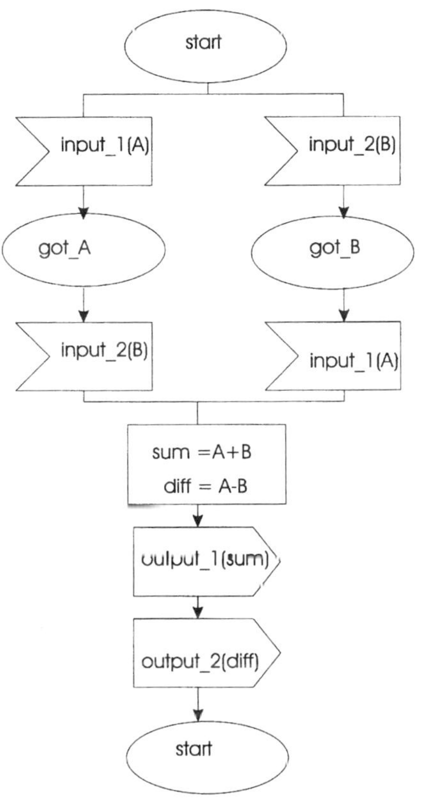

Figure 2.11 SDL specification of adder-subtracter.

Prolog

Prolog is a programming language based upon formal logic. It has been used as a specification language because of its elegant style and approach. The language basically is composed of a set of facts and rules defining relationships between objects, as well as queries to retrieve the relevant information from a logic program. This is helpful when a user is trying to specify the behavior or requirements for a complex system. The language is declarative as opposed to imperative. Rather than being concerned about how a given specification should be implemented, the language enables users to define what is being constructed.

This section will provide a brief overview of the language syntax which will make later chapters more readable for readers who are unfamiliar with the language. A complete treatment of Prolog is certainly not the goal of this section. Interested readers are urged to refer to Clocksin and Mellish (1984), and Covington, Nute, and Vellino (1988).

The fundamental elements in Prolog are called *terms* and consist of *atoms, variables, numbers, structures,* and *lists.* These are defined below.

Atoms. Atoms are names of objects and predicates in Prolog. They begin with a lowercase letter and can then contain any combination of letters, digits, or underscores. An atom enclosed in quotes can contain any type of alphanumeric character. There is also a set of reserved atoms, such as the plus sign "+" denoting an arithmetic addition.

Variables. A variable is denoted by any sequence of characters and digits that begins with a capitalized character or an underscore.

Numbers. Numbers are denoted in the usual way with any combination of digits.

Structures. A structure is defined as an atom or an atom followed by a parenthesized expression. Within the parentheses can be a set of one or more arguments separated by commas. An example of a structure is *workstation(BRAND, color_screen, '100 Mhz', multiprocessor).* Structures are used as building blocks for *facts, rules,* and *queries.* They can also be viewed as data structures. For example, the structure *workstation* described above may be used as a storage element for recording specific information regarding different brands of workstations.

Lists. A list is an ordered sequence of zero or more terms. The terms are contained within square brackets and separated by commas. For example, the following is a valid list: *[red, blue, yellow, noon,2].* A list has a *head* and *tail* associated with it. The head denotes the first element of a list, while the tail represents the remaining elements. For example, the head of the previous list is *red,* and the tail is *[blue, yellow, noon,2].* The tail can be a list or an empty list. The head and tail of a list can be extracted through the use of a "|" symbol. For example, a list can be expressed as *[Head|Tail].* Using the list which was previously shown, the variable *Head* will bind to *red* and *Tail* to *[blue, yellow, noon,2].*

Facts. Prolog utilizes terms to create *predicates, rules,* and *queries.* Relationships between objects are called *facts* or sometimes called predicates. For example, the fact

```
a_language(prolog).
```

may mean that *prolog is a language.* Here, the predicate *a language* attaches some kind of meaning to the argument *prolog.* A predicate can have any number of arguments to constitute a fact. The number of arguments is called the *arity* of the predicate. The following predicate has an arity of 2:

```
a_language(prolog, logic).
```

This statement may mean that *prolog is a language based on logic.* Note that the use of prepositions in interpreting the given fact is totally arbitrary; they are used only for convenience. The meaning could as easily have been given as *prolog is a language using logic.* Here, the implied relationship between the arguments and the predicate are important. Hence it is significant that the interpretation be consistent with other facts based on the same predicate.

Rules. Another basic statement in Prolog is a *rule,* which establishes dependencies among a set of facts. A rule is composed of a head and a body separated by a ":-". The head, which is a structure, is deemed to be true if the conjunction of structures that make up the body is satisfied. It is often convenient to interpret the ":-" symbol as the word *if.* For example,

```
is_a_building(X):-has_windows(X), has_doors(X), has_rooms(X).
```

can be read as: *X is a building if X has windows and X has doors and X has rooms.* Within the Prolog environment, facts would be stored to support the body. For example, a sample list of facts could be

```
has_windows(apartment).
has_windows(car).
```

```
has_doors(apartment).
has_doors(car).
has_rooms(apartment).
```

The rule would then be satisfied if *X is an apartment,* since *X* = *apartment* will match all the elements of the body [whereas *X* = *car* will not satisfy the last element, *has_rooms(X)*]. The ability of Prolog to satisfy the left-hand side of a rule (sometimes referred to as a goal) with a body is termed *unification.* This is a powerful concept. Unification is achieved by variable instantiation in the body of the rule. In the previous example, the variable *X* is temporarily instantiated as *apartment* and is tested against all three elements from the body. If all are satisfied, then the head (goal) is unified with the variable *X* instantiated as *apartment.*

The order in which facts and rules are entered into the knowledge base is important. Prolog will instantiate variables and test them according to the order in which they are entered. For example, if the first two entries in the knowledge base in our previous example were reversed to read

```
has_windows(car).
has_windows(apartment).
```

then Prolog would first instantiate *X* = *car,* since *has_windows(car)* is the first entry. However, when testing the last element of the body, *has_rooms(car),* Prolog would not find a fact in the knowledge base. Consequently, Prolog would *backtrack* to the point of failed variable instantiation and try another alternative for *X* (which in this case would be *apartment*).

Queries. Using rules and facts as its knowledge base, Prolog permits the query of the information. In the example given in the previous subsection, a query could be

```
is_a_building(NAME).
```

and a typical Prolog interpreter would reply with

```
NAME = apartment
```

indicating the response to the query. Hence queries can be used to extract pertinent information from rules and facts.

Specification in Prolog. Description of the adder-subtractor in Prolog is relatively intuitive. The system has two inputs and outputs and performs two mathematical operations. Hence the piece of code below would suffice to define the system:

```
adder_subtractor(Input_1, Input_2, Output_1, Output_2):-
Output_1 is (Input_1 + Input_2),
Output_2 is (Input_1−Input_2).
```

In general, Prolog provides a good platform for specifying systems because of its declarative properties. It is also a suitable language to quickly construct prototype or meta languages. These features of Prolog will be exploited in later chapters.

Simulation

A model by itself is of little use unless there is an environment in which it can be simulated and observed. A typical means of doing this is to use a simulation engine as the basis for observing events and recording outputs. Simulation is accepted as an integral part of the product life cycle in many industries. A simulation study typically progresses through a structured logical sequence of activities during the duration of the simulation life cycle. A graphical representation of the typical flow during a simulation project is given in Fig. 2.12. The boxes represent different steps in the cycle, and the arrows indicate the progression to the next possible step. Given the significant number of feedback loops evident in the process, automated steps to expedite the simulation cycle are highly desirable.

To enhance the productivity of the simulation process, the concept of a simulation environment evolved. This encompasses the underlying simulation language and support tools. Most of the nine steps identified in Fig. 2.12 can be addressed by a competent simulation environment. For example, model and program design can be aided by graphical representation tools, custom model editors, and model management utilities. All simulation environments utilize an internal language for simulating models. The ease of representation for a given language is a direct consequence of the level of expressiveness of the language with respect to a given application. This points to the notion of the ability of a language to "naturally model" a conceptualized system with the least amount of mental transformation and mapping on the part of the modeler.

Fundamentally, languages are capable of providing different forms of modeling power to the user. The language of choice would depend upon, apart from other factors, the application and the modeler's own background and preference. Modeling can be accomplished by numerous methods, including graphics, hierarchical construction, syntactical input, and interactive prompting.

Another basic capability of simulation languages is that they allow experimentation with models. The user can perturb the model and observe the changes at selected points in the model. Validation and

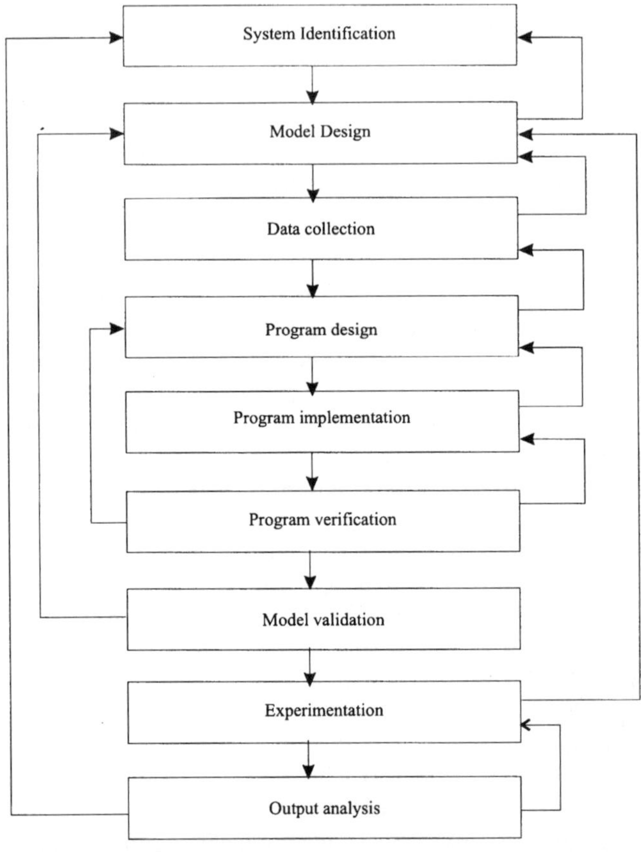

Figure 2.12 The simulation project cycle (*1995 IEEE*).

debugging are closely linked with these ideas, and different types of such capabilities are available within commercial environments.

The minimum constructs of a simulation environment can be classified as falling into two functional groups: host language dependencies and model specification.

Host language dependencies

A simulation environment provides an underlying computer program (implementation language) that supports the simulation activities. The performance and ease of representation of simulations are directly affected by the limitations of the underlying implementation language. To properly facilitate modeling and simulation activities, the language needs to support high-level data structures and operations. For example, if a language contains no dynamic memory management primitives, it may impose restrictions on the ability of the modeler to design large and complex interactions. The same limitation applies to strictly typed languages. That is, such a language may impose unnecessary requirements on the modeler that do not improve modeling concepts. Given the description of the model abstraction in the previous sections, an object-oriented language or a declarative language such as Prolog with object-oriented extensions would provide a good foundation.

Model specification

Support of model specification constitutes the second major group of activities a simulation environment must provide. These activities are the basis for the specification and use of models. The modeler should be free from implementation details of high-level constructs in the environment. That is, the environment should provide facilities for constructing, composing, and debugging model abstractions. In addition to support of the basic model abstraction, some other features of a simulation environment are desirable.

Formal basis. The validity of the modeling process for discrete event simulation can be established with a sound underlying basis. Many mathematically sound formalisms exist as options, such as SDL or Petri-nets. The formal basis that is chosen may very well determine the ease of validation or representation of a model and model base.

Modularity. Models are distinct objects that communicate with one another and the environment through their interface specifications. Hence, adherence to the concept and structure of the model abstraction implies modularity. The level of communication between models will be influenced by the attributes that are inherited from the host language.

Verification. An important and often neglected aspect of a simulation environment is the ability to test and verify the behavior of each submodel of a larger model. With the ability to test each model before building a larger and complex one, the modeler minimizes verification problems in the larger model. The interface specification of the model must permit viewing the behavior of a given model with respect to controllable stimuli. Decomposing a large model into its smaller constituents generally facilitates the management of the verification process.

Validation. Related to verification, validation of the model is an activity which tests the validity of the behavior of the model of the system under design. The test is with respect to design specifications or expected behavior of the system.

Experimentation. Experimentation is the phase that follows modeling in a design or analysis task. The purpose of experimentation is to learn more about the system under study by subjecting its model to various input sequences selected from the legitimate inputs of the model.

The process of constructing experiments resembles modeling, in that it can be considered modeling the environment of the system under evaluation. Consequently, additional models can be defined to provide the required stimuli to the modeled system. These models allow the controlled exercise of the system model through different simulation scenarios. When composed, models and their experimentation models constitute a closed world.

In addition to features supporting modeling, a simulation environment must also be able to support the design and conduct of experiments. Experiments can be derived from test specifications, load descriptions, or other means applicable to the design. The design of experiments can also be regarded as a modeling activity. The key issues surrounding good experimentation capabilities can be summarized as follows:

- Data or statistics gathering capabilities should provide the ability to monitor specific points of interaction in a model, such as inputs and outputs.
- Facilities for observing interactions between models and intercepting data (messages) are needed. The messages exchanged between models can help the user to relate the cause and effect of such exchanges.
- Library operations must be provided. These include provisions for storing a specific model and retrieving it in a structured and efficient manner.
- Model reuse dictates the ability to inherit or reuse the whole or a part of a given model. Modularity alone implies reusability at the model level. However, more support is required to ensure reuse of parts of a complex model at higher levels.
- Inheritance refers to the ability to extend the definition (or specification) of a model. Therefore, a new model can be generated from another by inheriting its specifications and incorporating additional features. Both of the last two elements are required in order to support model composition.

Domain-Directed Design

This chapter has presented some of the fundamental techniques and tools used in the design of complex systems. Many other tools are employed by designers; however, many of the ideas behind the ones presented here are typical of all of them. There are many more different tools that are used, but all exhibit some form of similarity. The tools are powerful aids to the designer, but many of them are disjoint and not well integrated into the overall development process, from specification to implementation. In particular, the knowledge gained during a simulation study is almost always discarded after completion of the simulation activity. The next generation of design environments will not be able to afford to do this, since such knowledge is a powerful piece of information for the design space exploration of alternative designs. Consequently, a cohesive design approach for complex systems which will integrate the various tools and formalism in a seamless way in order to enable users to create, experiment with, and synthesize a design from the outset is needed. To achieve this, two elements must be satisfied. First, a limitation on the scope of a design environment is required. This implies that a domain-specific or directed environment is preferable. Finally, an *architectural level* of abstraction must be supported. This criterion necessitates a powerful abstraction and design analysis mechanism.

A domain-directed approach provides many benefits. The current state of the art in simulation provides ways of organizing and using the model knowledge. Modeling is usually not a stand-alone activity. It is typically a part of a design or analysis project. Therefore, the models may be used in many different ways. For example, a model may need to be stored and checked against other models for equivalence, and so forth. Consequently, there must be higher-level constructs to manage such functions in an efficient way, which we call knowledge management.

Knowledge management is a crucial step in creating a domain-specific environment for complex systems. Knowledge about a given structure or classes of models can be employed by the environment to assist designers in making more educated choices. There are, however, clear limitations. The knowledge base population, or the number of models contained in the environment, directly affects the ability of the environment to aid the designer. To make the knowledge relevant, the information must be restricted to a specific application domain. For example, the telecommunications domain may wish to make available features for easy modeling of transmission lines, digital switches, or modulators. Most of these requirements would not be applicable to another domain, such as flexible manufacturing systems. Hence separate requirements for the structure and use of the knowledge would be needed for the two domains.

Such domain-directed support also implies higher-level system-level design tools and integration. These can be automated design documentation, configuration management, replication, synthesis, or other automated components for design construction. A typical example is a graphics- and text-based tool that identifies model compositions and synthesizes the actual simulation or specification model mainly from the graphical descriptions. Another example is a mechanism for organizing models in a structured manner, whereby model interconnection can be managed by a knowledge management and library support system. When a complex, reusable model is requested from the library by a parent model, the environment can determine and create the necessary model interconnections and resources to effectively configure the model to function with the parent.

The types of operations within domain-oriented support are valuable to any design environment. The use of this along with the modeling, representation, and simulation concepts is the basis for *architectural-level design.* If we revisit our original discussion on art versus science, architectural-level design somewhat attempts to bridge the gap between the two (refer to Fig. 2.13). Compared to traditional design environments (embedded in one of the product development processes), the architectural model permits a larger degree of artistic control, while maintaining a firm process-oriented basis. This permits

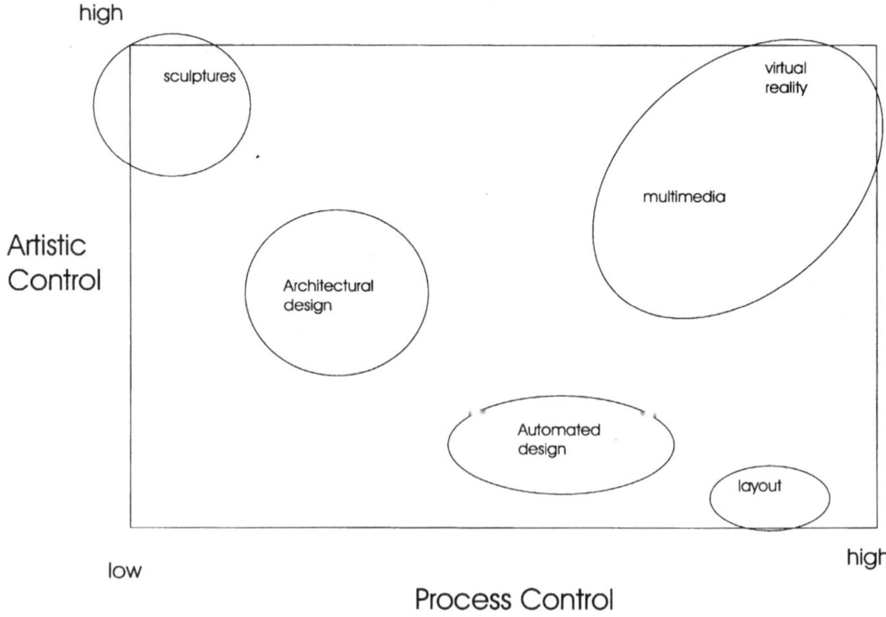

Figure 2.13 Process versus artistic control of sample design environments.

the construction of concrete design implementations from abstract concepts from design specifications and/or requirements.

This chapter has provided some of the background necessary to explore the design of complex systems. The following chapters will illustrate how these ideas can be applied within the context of a practical design environment.

Chapter

3

Architectural Design

The earlier chapters have defined the architectural level of abstraction as a level of realization just after a conceptual description or specification of a system. This chapter will examine the common issues prevalent in the design of systems at this level of abstraction. The umbrella term *architectural design* will be used to denote all design activities undertaken at this level.

Introduction

Architectural design can be viewed as a transformation function, represented in Fig. 3.1. The function accepts design specifications, design requirements, and design constraints as input. For the purposes of

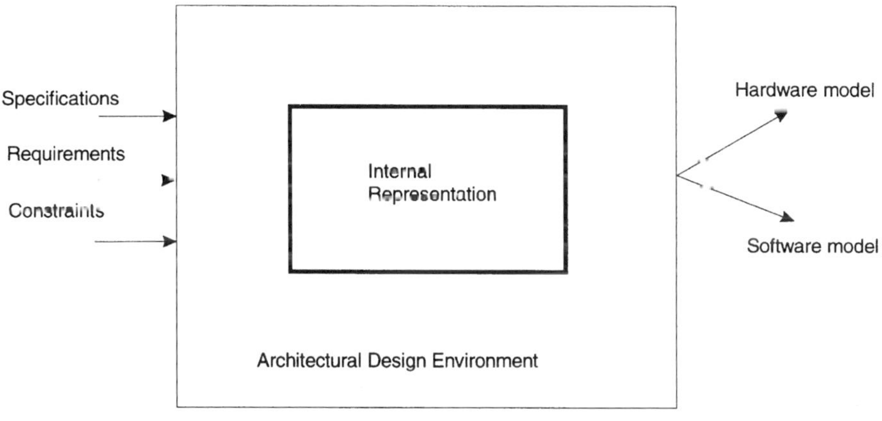

Figure 3.1 The function of architectural design.

architectural design, the difference between design specifications and requirements will be defined as follows:

- *Specifications* constitute a formal definition of the system design. In general (but not always), the notation will allude to *what* the architecture may consist of (such as the protocol it is to use).
- *Requirements* represent an informal or interpretable description of the system. Therefore, it is the designer's interpretation which will greatly flavor the initial architecture that is selected.

Through a range of activities, the function produces as output a hardware model and software model of a design that is consistent with the criteria presented at the inputs. The nature of architectural design implies a hierarchical approach to the design of a system, progressing from high-level system detail to a low-level description. However, there is a clear distinction here between the inputs that would typically be found when dealing with lower levels of design abstraction. First, there is the existence of both inputs that are described in a formal notation (the specifications) and some that are informal (requirements and possibly constraints). Hence the function must be able to deal with these possibilities. Ideally, the transformation function would perform an optimization such that the resulting output is the best solution. Practically, this is rarely possible. It is more likely to be a solution derived from existing information on alternative designs.

An essential ingredient in making an architectural design environment work is the use of an internal medium for capturing the input. The medium is a representation akin to specification languages (in reality, the medium must possess additional constructs which are not found in the typical role description of specification languages). Examples of specification languages were presented in the preceding chapter. Although the languages may be adept at handling formal descriptions, additional means for capturing the requirements and constraints in some intermediate form are needed.

The language (or group of languages), which will provisionally be referred to as an *intermediate language,* needs to accommodate both the formal and informal representations. This is not a simple task. System designers have typically utilized many different languages to capture relevant aspects of a system. It is also generally agreed that one *unified* language or methodology is not capable of representing systems at all levels of design abstraction. It is also observed that as the abstraction level increases, the analysis methods used in tools shift from simulation-oriented to formally based methods (Ward 1985). Hence it is difficult to use an existing paradigm to address design concerns across a broad range of activities at the architectural level of design.

All this implies that the intermediate representation supported by the environment must undergo some transformation during the design activities to change the vague semantics into less ambiguous ones. This transformation could be made possible with some constructive interjection by the designer and resources allocated to the environment.

Architectural Issues

Architectural-level issues are prevalent during the design phase of product development. In particular, the implications of this level of abstraction are observed very early in the design trajectory. The various activities in a design phase that are typically pursued within the context of architectural design are shown in Fig. 3.2. The three major

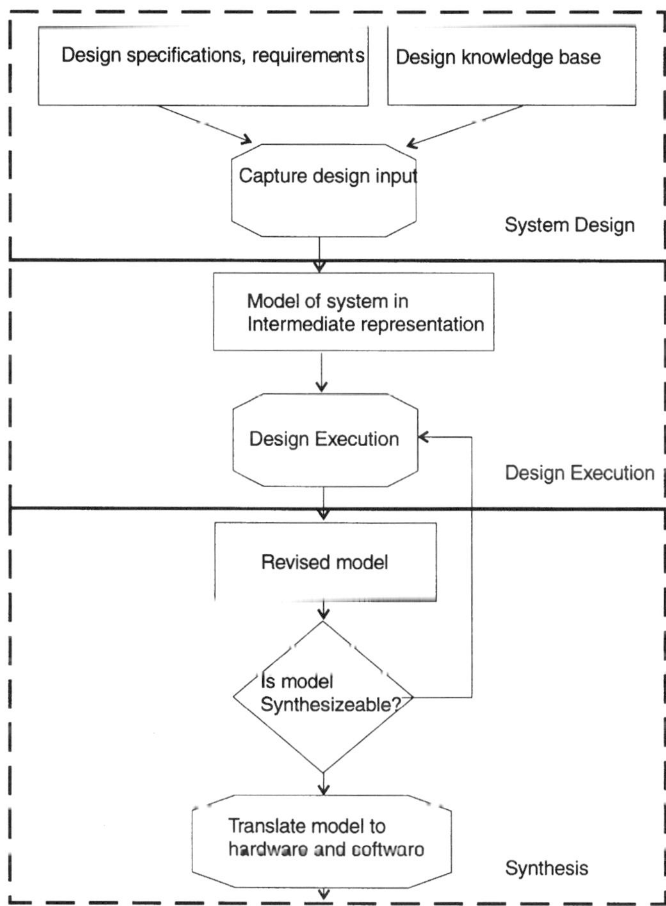

Figure 3.2 The architectural design flow.

design activities that were alluded to in Chap. 1 are broken into more detailed activities to highlight the progression of the design. The design flow (the inertia of activities resulting in some outcome) is from design capture to synthesis.

The system design activity is the front end to the design environment. It provides the means for capturing user-supplied design specifications and requirements in a form suitable for the design environment. The input can vary depending upon the domain of application. For example, SDL may be an appropriate input for designers of protocol-based systems. This specification would need to be captured within the intermediate language of the environment and conditioned to be used within the environment. The execution-related activities are centered around design space exploration. This implies executing different design architectures to eventually derive an acceptable design. The execution phase may undergo several iterations that are part of a model refinement exercise. This requires the support of a model library capable of providing refinements that can make the intermediate model a more detailed one which is amenable to synthesis [Booch 1991]. The synthesis activity uses the final model to generate hardware and software representations suitable for use by other design automation tools.

The three major activities of design are broken down into their inputs and outputs in the following sections. In addition, whenever applicable, basic requirements for achieving the transformation from the input to the output of each activity are provided.

System design

The system design activity is generally the starting point within the design phase of a product life cycle. The activity involves the design capture of specifications in a model which can then be executed. Hence the language requirements at this stage encompass modeling capabilities. A model represents an abstraction of a domain under study, and a large body of knowledge exists with respect to this abstraction. In the context of architectural-level design, an abstraction is closely linked to specification capture, reusability, and object-oriented features. However, from the model perspective, it can be seen as representing a set of software or hardware functionalities. This transparency lies in the fact that the behavior as expressed by the model can be viewed as a set of communicating processes or states. Hence at this high level of description there is no direct correspondence to the underlying hardware or software nuances.

When modeling at the architectural level, the user also adopts a global viewpoint, where not only the behavior of the system under

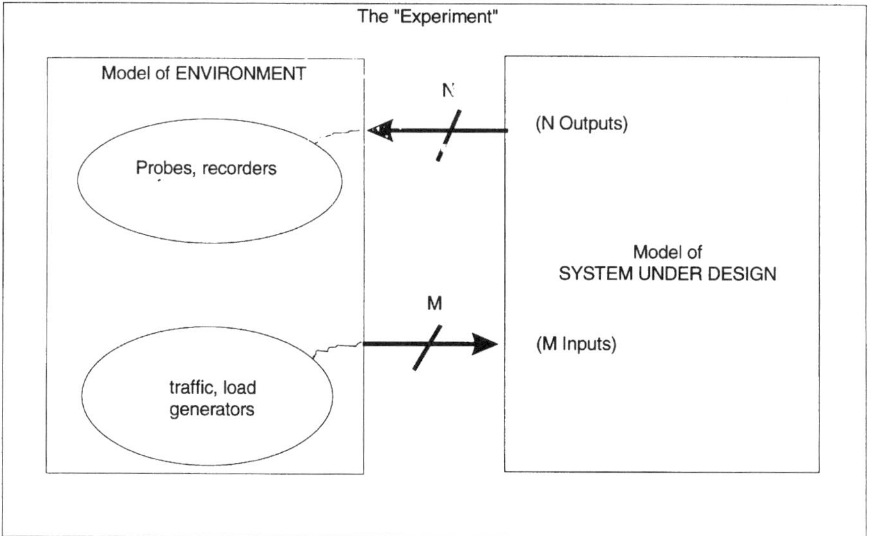

Figure 3.3 The system under design and its surroundings.

design but also its immediate environment is relevant (refer to Fig. 3.3). The environment, in this sense, constitutes the generators of all the inputs to the system under design as well as any *probes* needed to record or observe the outputs.

The notion of treating the system under design and its environment as two interacting models is a well-known system design approach (see Zeigler 1984 and 1993). The beauty of this approach is that the two systems can be modeled, simulated, and analyzed using the same underlying formalism. Hence, the environment and system under design are, for all intents and purposes, entities that are constructed using the notions of a *model* as introduced in the preceding chapter. Therefore, the entire experiment, which is a result of the two models interacting, is viewed as a single *composite* model.

Given that the experiment is a composite model, there are important ramifications for the architectural level designer. First, a library composed of generic environment entities can be maintained. For example, traffic distribution generators can be created, tested, and maintained as a library model. They can then be reused for any future design. The generic entities also lend themselves to custom-designed traffic generators, where an existing model of a traffic generator can be used as a template for constructing a variant.

Similarly, the output probes to the system can also be created as library models. Hence a library of statistical probes and recorders can be used by modelers to observe and analyze the output of the system under design.

Inputs to system design. A variety of inputs are possible at this stage. These represent the system using a multiplicity of notations and descriptions, working at possibly different levels of abstraction. The inputs are:

- *Formal specifications.* Specification capture has been an evolving discipline in system engineering. The representations presented in Chap. 2 can also be used as input at this level. Today's system engineer will find many different mechanisms for design capture (Rattray 1989). Of these, one with an underlying formal basis is desirable to ensure consistency in models (Jacobson 1992). The sample languages presented in the previous chapter comply with this norm in one form or another. For example, SDL utilizes an extended-state machine-based formalism, and Lotos a process algebra–based one (Milner 1980; Hoare 1985). Both have been developed primarily to specify telecommunication systems. Hardware-oriented formalisms can also be used, such as HOP (Gopalakrishnan 1989), based upon a functional programming paradigm. More general formalisms such as Petri-nets (and their extensions) (Peterson 1981), temporal logic (Moszkowski 1985), and queueing networks (Kleinrock 1976) can be employed to capture and analyze different aspects of a design. Additional paradigms can be found in Gupta (1992).
- *Informal requirements.* These are requirements that are based upon bounds such as response time (average, minimum, and maximum), memory space requirements, or cost estimations.
- *Unknown elements.* These are components that are not concrete in terms of specifications. For example, a local area network interface definition in a system may not be known and can possibly use a variety of existing industry-standard protocols. Hence this must be reflected somehow at this stage of design.
- *Multilevel abstractions.* Some aspects of the input can be well defined, while others are not so clear, because some elements of the specification are given in greater detail. For example, formal specifications may exist for some aspects of the design and informal text for others. The variation in the level of detail can be significant and extremely cumbersome to handle with a specific specification language. Hence the design environment must permit the use of different levels of detail and formalisms in the model.

Outputs of system design. The output of the system design activity is an *internal representation* which reflects the behavior and structure imposed by the inputs. To facilitate addressing the different input possibilities, the internal representation language may need object-oriented capabilities or extensions. This is to minimize the large amount of complexity in transactions between the diverse model constructs. It should also be noted that models defined in the internal representation can also be used as inputs if so desired (especially if the internal representation is a well-defined formalism).

Design execution

After a suitable model has been defined, the language must provide support for experimentation and analysis. The activities encompassed by experimentation and analysis will be referred to as *design execution* (or simply *execution*). These activities are crucial aids for a designer in exploring a given design space, making appropriate trade-offs, and partitioning the design to different hardware/software configurations. Such activities can be supported through design simulators and formal methods. The input to the design execution activity is the model as described by an internal representation. The output of the activity is a transformed model. The transformation of the model is based upon many different elements (refer to Fig. 3.4). These are the analysis methods applied to the model along with any of the other techniques used in design described in Chap. 2.

The derived or transformed model is achieved through multiple intermediate refinements. These incremental changes to the original model are a consequence of the interaction of the analysis, execution engine, and decision entities of the environment. The user-controlled

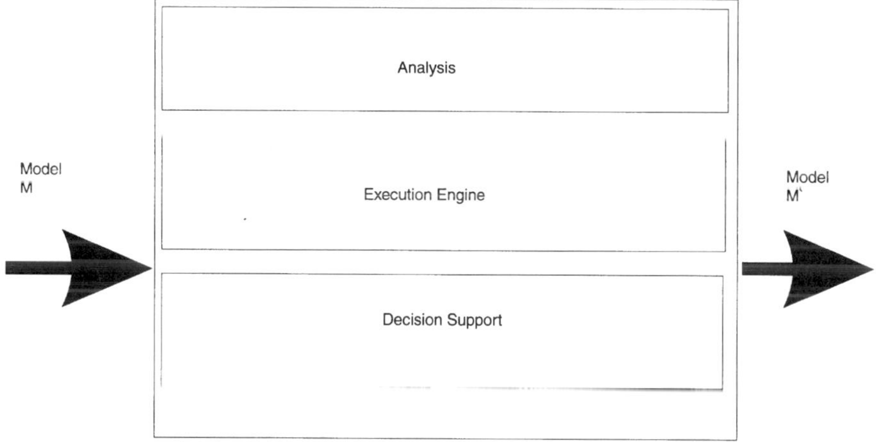

Figure 3.4 Design execution.

incremental enhancements to the model eventually converge to a detailed model which can be used by automated design aids to generate the hardware and software components of the system. The resulting design is not necessarily optimal (a very difficult problem in itself), but it is a solution within the given set of specifications. The three components of design execution are described below.

Analysis. Formal methods refer to the application of mathematical techniques to prove properties of a system model, such as existence of deadlocks. Although using formal analysis in all aspects of design is highly desirable, current methods make it very difficult to formally analyze large systems defined at detailed levels. The number of possible states of the system is just too large to manage. Additional problems with current formal verification approaches also persist. Formal techniques apply to the verification of a model of the system, not the system itself. The relevance of the verification is always dependent upon how well the model represents all aspects of the system. Considerable effort can be expended to formally verify a model, yet the model may not adequately describe the behavior of the system in sufficient detail to address all design queries. In reality, the amount of time required to formally verify complex systems is usually unacceptably long.

Execution engine. Simulation, where a given model is executed through predefined test scenarios depicting typical (and worst-case) operations that the system may encounter, is another technique for understanding system behavior. For large systems, the scenarios cannot be exhaustive and provide a limited degree of confidence in the system. Hence, the validity of simulation results is heavily dependent upon the assumptions the designer provides. Simulation can be employed in cases where insufficient detail is available for a formal method to be used effectively in the analysis. The simulation engine is also well suited for time-stepping through the model to test different alternatives and refining it. In this sense, it can also be viewed as the *execution engine* for architectural design.

Simulation is widely deployed as a major design aid. There are many general-purpose simulators, such as GPSS (Schriber 1974), SLAM-II (Pritsker 1986), and SIMSCRIPT (CACI 1987), that can provide reasonable queuing-based system simulation. Other systems, such as Designers Workbench, also provide analysis support for small systems (Thomas 1991). Large and complex simulation experiments that require parallel or distributed processors and techniques like Time-Warp (Jefferson 1983) have been utilized, generating speedups in most cases (Reed 1987). Petri-net–based simulators such as Voltaire (Parent 1991) and Loopn (Lakos 1991) can also be used for concurrent applications.

With the assortment of simulation languages, model reuse becomes very difficult. Model interchange between different tools is generally not supported, and no standard currently exists for these languages that would permit such an operation, although work is under way within international standards working groups to alleviate this problem and define requirements for a standard simulation environment (Tanir 1994c).

The problems of simulation language users are compounded by the fact that the terminology is also not uniform across product lines. The simulation community spans many disciplines, and depending upon the field of use, simulation terminology can vary or some terms (such as *model* and *entity*) can be loosely utilized. Hence it is useful to define basic terms that are generally accepted within the discrete-event simulation milieu.

- A *simulation model* is defined as a representation of a conceptualized system implemented within a simulation language.
- An *entity* is any component of a system that requires explicit representation in a model. Entities possess attributes denoting specific properties.
- A *state* of a system is defined as all the variables that are required to uniquely define the system at any point in time.
- An *event* is an occurrence of an operation at an isolated point in time which may result in a change of the system state.
- An *activity* is an operation over a time period of finite length that causes a change in the system state.

Simulation languages employ one or more of the three prevalent modeling views: event scheduling, process interaction, and activity scanning. These are commonly defined in the literature as follows:

- *Event scheduling* is a view in which the simulated system is constructed by observing the interaction of events and their respective effects on the system state.
- *Process interaction* is a view in which the system is seen as a set of overlapping activities, causing events as they begin or end; the activities form related groups, which are processes.
- *Activity scanning* involves defining the conditions necessary to schedule each activity in a system.

An example of the use of these terms is next presented. If we wish to simulate a simple bank with customers wishing to either deposit or withdraw money, the definitions could be interpreted as follows:

System:	*A bank*
Entities:	*Customers (attribute: bank balance), Tellers (attribute: average service time)*
Activities:	*Depositing, withdrawing, waiting*
States:	*Teller idle, teller busy, customer waiting*
Event:	*Customer arrival, customer departure*

The above notions form the foundation for most general-purpose simulation languages. However, further utility can be added to the basic languages by incorporating decision support capacity in the underlying language.

Decision support. Components to aid the designer in completing the transformation of the initial model to a well-defined one are also required. These are loosely termed *decision support tools.* As the execution engine moves the model through simulated time, various tools are needed to complete the successful execution of the model. These tools form the basis of the *cohesive design* view and the requirements previously stipulated for a simulation environment (refer to Chap. 2). At the architectural level of design, these are not just amiable features of an environment, but are necessary to produce a usable design.

A decision support system implies the integration of two components of the design environment: a library support system and a rule-based engine.

- The *library support system* provides many essential elements to the designer. It can facilitate the configuration of the executable model by providing reusable models from a model base. Hence, it is desirable for the library to have powerful storage, retrieval, and management features. The library should also maintain a class of models. This permits the environment to interject alternatives during execution errors resulting from inadequate model choices. Such conditions can typically occur if the modeled subcomponent does not perform within the initial specifications.
- A *rule-based engine* is needed to permit the environment to compare the system under design with past design knowledge. A rule support system can also facilitate the configuration of models extracted from a model library. For example, if a given library model can indicate how it needs to be structurally and behaviorally instantiated within a given system model, the process can be automated by the system, permitting the construction of generic models.

Synthesis

Model execution activities are repeated, refining the model, until a final model is derived which meets the designer's expectations and which can also be synthesized. In general, synthesis implies taking a set of behaviors, constraints, and goals and generating a suitable structure that can implement the behavior while satisfying the constraints and goals.

Software synthesis at this point requires significant research to automate an accepted high-level specification within computer-aided software environments and methodologies. The hardware field has seen a proliferation of design languages (along with the IEEE standard VHDL) which can be used as potential target languages for hardware synthesis. The hardware design community has actively defined many design description languages predating the standard VHDL; CASCADE (Borrione 1993), CONLAN (Piloty 1980), ELLA (Morrison 1993), and Verilog (Verilog 1991) are examples of the languages that are available. Synthesis has been one of the most important applications of hardware description languages. Initially targeted to circuit-level synthesis, tools and methodologies have evolved to synthesize designs from the algorithmic level.

Although highly significant, synthesis has played a secondary role to the design language definition. For example, only restricted subsets of VHDL and Verilog are suitable for synthesis. These languages have powerful constructs for simulation of hardware; however (in part because of their low-level modeling features), automatic refinement of models has proven to be elusive at the very high levels of abstraction used within the architectural level of design.

Within the realm of the architectural level, synthesis has some further connotations: It must support design partitioning, timing transformations, and translation of behavior.

Partitioning. Traditional synthesis techniques and tools which are applied at lower levels of design abstraction concentrate on either hardware or software. Partitioning at these lower levels of abstraction is concerned more with resource allocation and data flow optimization. At the architectural level, partitioning implies the identification of the software and hardware components, since they are initially indistinguishable. This is not a trivial task and requires some heuristics and optimization criteria.

The first step is to identify distinct components in the design (refer to Fig. 3.5). These will be termed *system* models and constitute the elements whose interfaces are well defined and could be synthesized by themselves. Although this step is very difficult to perform with an algorithm, it is natural through designer intervention and heuristics. A heuristic which can be effective in most cases attempts to determine if a model element is an instance from a library—and, if so, whether it is the parent object. A parent object typically possesses clearly defined interfaces (or rules and methods for defining the inter-

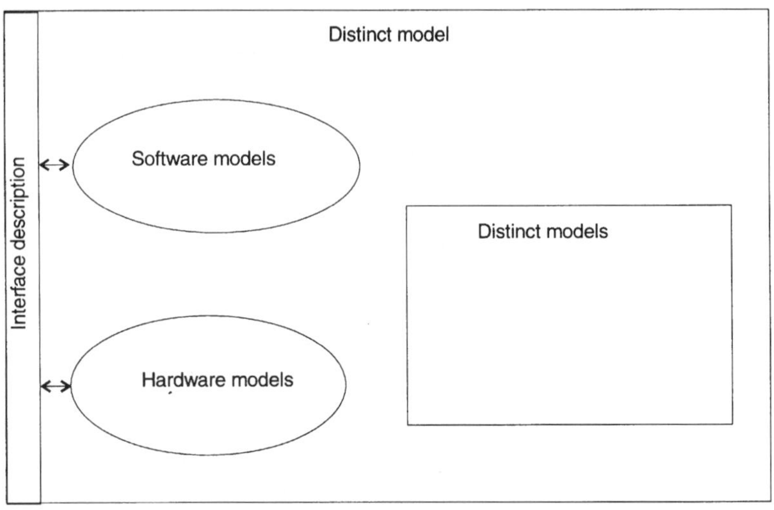

Figure 3.5 Model partitioning elements.

faces). The designer naturally should have the final say in the identification.

The second step in partitioning is to take each of the distinct system elements and breakdown and tag its constituent components as software or as hardware or as another system element. This step is feasible mostly through designer intervention. However, again some heuristics can be employed. For example, the frequency and complexity of interactions between respective elements can provide some indication as to their functionality. If a given model is seen to be communicating with another model above a predefined threshold frequency, then it may make sense to group the two models as either software or hardware. This may reduce timing delays but could increase costs. The final decision really falls on the shoulders of the designer.

Timing. Timing issues must be resolved from a high level of abstraction to a relatively low level. This implies transforming timing information in the form of message-passing attributes into a form defined in terms of hardware timing diagrams or software timing constraints. The former case raises some interesting issues. Assume that two architectural-level submodels are communicating by exchanging high-level messages. The term *high-level* here denotes a level of abstraction which is higher than a representation of electronic signals on a wire (or timing relationships in a timing diagram). How then can a translation be made from the message(s) to the timing diagrams?

For example, if a message is equivalent to "reset contents of output buffer" (see Fig. 3.6), what is the equivalent timing notation needed to

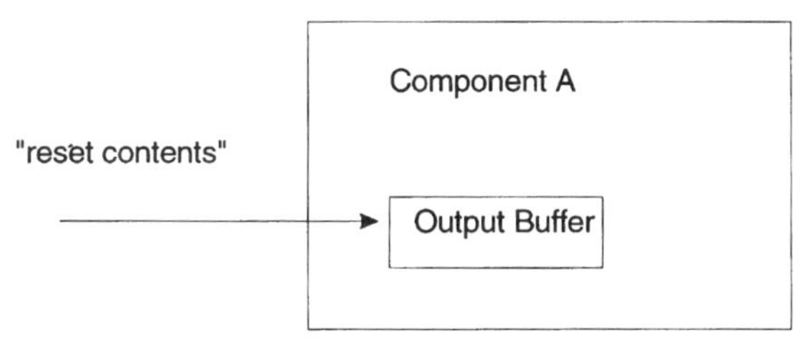

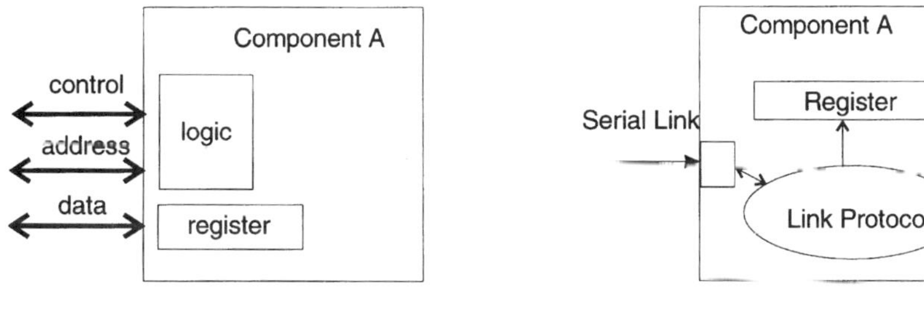

Possible timing synthesis option A. Possible timing synthesis option B.

Figure 3.6 Timing synthesis ambiguities.

support this at a hardware level? One possibility is to generate a set of wires to transfer the data and another set to facilitate the control. Another possibility is that the message is really part of a high-level protocol over a transport facility. In this case, the protocol semantics must be captured by either software or hardware.

The solution is not apparent and requires assistance from the designer as well as possible inference from the environment's knowledge base to derive a suitable design.

Translation. Finally, there is a translation component from the high level (captured within an intermediate representation) to the lower representational form. The translation must be consistent with the knowledge gained during design execution. Hence design modifications introduced during design execution must be recorded and reflected within the translation.

To support translation, the environment must provide an added capability that is not traditionally part of a simulation or synthesis environment. This is the capacity to record and store design changes made during the design execution phase. The records must provide the behavioral information as well as structural modifications. Conventional simulation environments have provisions for observing the execution and capturing statistical information. However, the underlying simulation model in these environments is static in both structure and behavior. In the architectural design world, the model is viewed as a dynamic entity, changing its behavior and structure during the course of design execution. In this sense, an architectural-level model assumes a design history which reflects the different roles it has played.

Intermediate Language: The Melting Pot

The issues reflected in this chapter lead to the conclusion that different paradigms are applicable at different stages of design. Furthermore, there are additional needs dictated at the architectural level, making it difficult to utilize an existing general-purpose specification or design language throughout the design process. However, it is clear that some uniform underlying language is required to facilitate all the activities that were alluded to previously. Consequently, a possible solution to support architectural design is to utilize an *intermediate* language within the environment to obtain the various input and output forms. Such a language would support inputs in the form of specifications (possibly defined in an existing specification language) and output a lower-level synthesized design in a usable representation (such as VHDL). The intermediate language would also require modeling capabilities for high-level descriptions to facilitate specification input, as well as design exploration and synthesis constructs to produce the outputs.

Design languages exist for a myriad of different applications. Unlike register-transfer/circuit-level design, the architectural level blurs the distinction between software and hardware, and therefore requires elements common to languages applied in both hardware and software.

In the reviewing of the architectural issues, many intermediate language concerns were highlighted in this chapter. The language is utilized during all three phases of architectural design, and as such, the respective activities for the phases reflect the requirements for the language. To summarize, the basic requirements for an intermediate language to support architectural design are:

- It must be able to capture formal specifications. This implies that the language should have formal foundations, or that a direct translation from the formal specification to a subset of the language should be possible.
- It must be able to capture informal and partial specifications. This implies that the language can permit the representation of incomplete structures, such as missing interface descriptions.

- It must support high-level communication paradigms (typical of software), such as message passing.
- It must be able to represent low-level communication entities typically found in hardware description languages.
- It must be able to distinguish between software and hardware elements.
- It must provide for the dynamic alteration of a model's structure and behavior as required by the designer or environment. This implies that the interface descriptions of the models can be altered on the fly at run time.

The essence of these requirements is shown in Table 3.1. The table also indicates the applicability of the various representation formalisms discussed in previous chapters. The table does not provide an indication of the amount of effort needed to achieve the requirements. For example, SDL naturally provides message passing–based communication, while Prolog will require additional programming to establish this desired feature. In general, meeting the requirements with Prolog will require additional effort (compared to the other representational forms). However, the utility of Prolog in filling the voids where the other forms are not suitable is evident. In particular, the dynamic modification requirement is not addressed by any of the other formalisms.

The table also depicts the complementary aspect of the formal specification formalisms such as SDL and predicate nets and a declarative language such as Prolog. In particular, the first two are naturally well suited for describing the known structure and behavior of a system, while the latter provides a vehicle for dealing with the unknown elements. Hence, a useful role for Prolog can be during design execution, when ambiguities in the design must be manipulated (such as unspecified or unknown interconnections between models). This rationale leads to the idea of using Prolog as a basis for the intermediate language.

TABLE 3.1 Intermediate Language Requirements versus Representation Forms

Requirements	SDL	Predicate nets	VHDL	Prolog
Formal specifications	Yes	Yes	No	No
Informal specifications/ambiguity	No	No	No	Yes
Support message communication	Yes	Yes	Limited	Yes
Support signals (wires)	No	No	Yes	Yes
Distinguish hardware and software	No	Limited	Limited	Yes
Dynamic modification of structure	No	No	No	Yes

Example

Before proceeding further with the intermediate language details, it is important to appreciate the approach that is usually taken during the architectural-level abstraction. An example illustrating the level of abstraction would be useful. This section introduces the design of a significantly complex telecommunication system. The example will be used progressively in later chapters to illustrate the hardware and software modeling issues that will arise at the architectural level of design. In this section, the design problem will be stated. The objective is to create an awareness of the issues that a typical architectural designer may face.

Description of problem

The example that will be studied is based upon the design of a traditional digital voice switch providing plain old telephone service (POTS). From a designer's viewpoint, it is useful to envision the design as that of a client and service provider relationship. That is, the switch will provide a specific service to the telephone user (the client). Given the title of the problem, system requirements (based upon a typical basic switch platform) for our POTS service provider are defined below. The description is in a format that may typically be provided to a design team as part of the system requirements documentation.

System requirements for voice switch. The goal of this design is to create a new, inexpensive digital switch to serve our customers in remotely populated areas. The basic functionality of a digital telephone switch system (henceforth referred to as DTSS) is to accept dialed requests originating from a basic telephone and establish a connection to the destination telephone. To accomplish this, the DTSS must provide the following services:

- Detect an incoming call (off-hook) from the source.
- Provide a dial tone to the source.
- Collect digits as transmitted from the source.
- Translate digits to determine the destination's physical connection to the switch.
- Select the destination's port on the switch.
- Establish a connection between the source and the destination.
- Send a ring tone to the destination.
- Send a ring tone to the sender.

- Send a busy tone to the sender.
- Detect a line busy signal at the destination.
- Detect an off-hook signal at the destination.
- Detect a disconnect (on-hook) and clean up all data structures.

The service must be provided within a reasonable amount of time. A dial tone must be provided to the sender's telephone within 1 second after an off-hook is detected at the phone. A connection must be established for local calls within 2 seconds on the average after the last digit is dialed, with a maximum of a 4.5-second wait time.

The hypothetical DTSS is expected to serve small residential areas and must have connectivity to a central office to permit long-distance communications. The typical configuration will provide service to 100 concurrent lines, with interconnection to 1000 customers. Telephone interface cards (TI-500) will be used to provide the analog interconnection to each telephone. These cards detect dial tones, busy signals, digits, and on-hook/off-hook signals originating from the telephones. It is also desirable to use the existing voice backplane BPC-1000 for the design of the voice bus so that it is compatible with other product lines.

Identification of design inputs

The first challenge in a given design problem is to understand the design requirements by identifying the inputs to the design. In this case, there is a mixture of various levels of detail. The functional description of the service provider is at a sufficient level of detail to reflect the sequence of operations that the switch must perform to provide its service for a customer. The level of detail could just as easily have been expressed as an SDL specification to ensure a more formal description of the service.

Other design constraints have also been presented in the problem. Performance-related requirements place bounds on how fast the service must be provided (the dial tone and circuit connection). A capacity requirement is indicated by the minimum number of customers that must be supported. A need to reuse an existing component is identified. In addition, there are vague or ambiguous requirements. A need to connect the switch to a central office has been indicated, but there are no details as to how the connection would be made or at what capacity.

There are also domain-related constraints that are generally implied or taken for granted. For example, telephone connectivity is established through serial lines using *pulse-coded modulation*. Connectivity between switches is established through time-multiplexed standards to provide for higher-capacity lines. The serial time-multiplexed lines are given various specifications and conventions within the telecommunication domain. For example, the number of time slots available per line can vary depending on the transmission rate. The standard for digital transmission in North America is DS1. DS1 carries PCM signals in a 193-bit frame composed of 24 (8-bit) channels (for voice samples) and a single framing bit. Transmitted at a rate of 1.54 Mbits/s, this ensures that a frame is transmitted every 125 μs. For telephone-grade quality of service, a 4-kHz voice needs to be sampled at 8000 times per second—which implies a sample (byte) every 125 μs. Hence a DS1 transmission facility can carry up to 24 separate voice samples across a single line.

Starting architecture

With any new design, the most difficult place to begin is determining the initial architecture. Sometimes there is no option but to make use of legacy systems for compatibility reasons, but there is usually a great deal of flexibility concerning choices in the internal design. In this case, some components have been defined as being applicable to this design. These are the telephone interface cards and a backplane bus design.

Applying a hierarchical approach to design, a particular architectural-level description of a DTSS can be thought of as being composed of three major high-level elements: a switch element, an interface element, and a service provisioning element (all shown in Fig. 3.7). The top-down approach decomposes the major elements of the design into manageable subelements. The rationale in this case is straightfor-

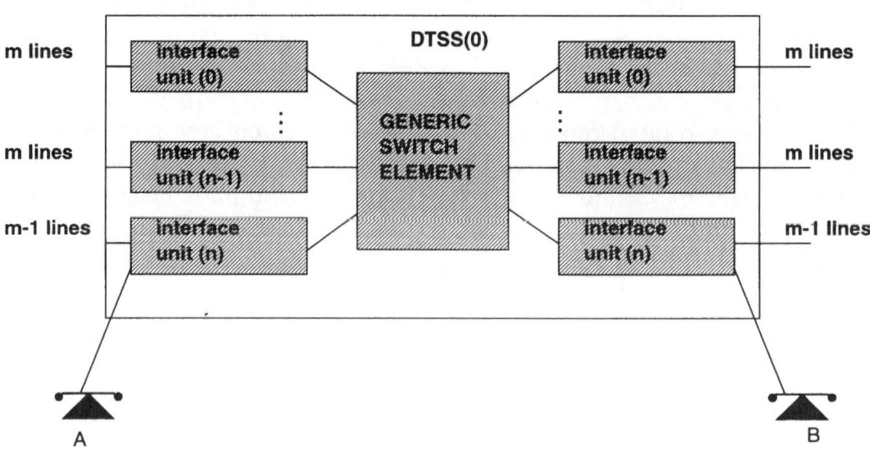

Figure 3.7 Architectural view of design example.

ward. The switch acts as a multiplexer and demultiplexer at its inputs and outputs, while it must provide a switching fabric for maintaining the connections. Hence, the interface element supports the multiplexing/demultiplexing functionality, the switch element provides the switching medium, and the service provisioning element ensures that connections are managed in the proper form.

Given that the switch must communicate with other switches in real time using a time-multiplexed standard transmission rate, the switching fabric has to process the time-multiplexed voice signals. The operation of the DTSS is as follows: An interface unit digitally samples m pulse-code–modulated (PCM) telephone calls and time-multiplexes them toward the switching element at an allocated time slot s. The switching element then transfers the time-slotted data to different outgoing lines and slots. The interconnection between the interface unit and the switch fabric is through serial time-multiplexed lines, so the fabric does not need to be burdened with any translation of the information.

Generally, the number of time slots s available between the interface units and the switching element is $m \geq s$. This is because telephones connected to an interface module are not all active at once. However, if more than s calls occur at the same time, then blocking of some calls will occur (a fast busy signal is sent by the network to the telephone). To provide for a flexible design that can be expanded, the number of interface units can be increased to accommodate more traffic. The number of interface units n and time slots to use are design issues based upon assumed traffic load criteria for the network.

The service providing (telephone services) element can be located within the switching element. The function of this element is to provide the call-setup support needed to establish a two-way telephone service. This element is assumed to be software. This is desired, since it allows additional voice features to be added to the switch in the future via software rather than by building custom hardware.

The switching element is the most real time dependent component of the design. Its function is to take a time-slotted incoming digital stream of voice samples (along incoming serial lines) and switch them to potentially different output lines (space switching) at different time slots (time switching).

As a starting point, the DS1 transmission standard will be assumed to be used between the interface units and the switch element; however, the design will be flexible so as to allow different numbers of channels, always guaranteeing the minimum requirement of 125 μs per frame.

Chapter

4

System Design Support

In this chapter, a complete description of an intermediate language to support architectural design will be presented. Implementation of this language as a meta-language will also be addressed. The language is an example of how architectural modeling needs can be defined linguistically and syntactically. The chapter will focus on the design support functionality of the language; hence most of the issues covered will be related to the representation and capture of design. The early sections of the chapter will formally describe the language. The DTSS example will be revisited at the end of the chapter.

Introduction

This section will introduce the design specification language (DSL), a description of an intermediate language suitable for use in an architectural design environment. DSL fulfills the mandate of an intermediate language by capturing specifications with the use of abstractions and is free from the rigid disciplines of simulator-oriented hardware languages. In addition, since the language must bridge the abstraction gap between high-level system design notions and low-level hardware descriptors, it must also have the flexibility to redefine and alter its model interfaces during design exploration.

The required features of DSL made Prolog a suitable implementation environment for it. It should be noted that DSL is a meta-language in the sense that it is based upon logic programming semantics, utilizing built-in predicates to define its own constructs. A novel feature of DSL is its ability to capture high-level specifications within a reusable model framework and, with the aid of a library support system, to refine the model to a state where it can be synthesized into a lower-level executable representation such as VHDL.

DSL inherits the typelessness of Prolog to provide facilities for powerful abstraction and data manipulation. The language is also designed to have a corresponding graphical notation and address the different architectural design roles it must play.

Notation

The following sections will present DSL modeling primitives required for design capture, addressing the design input concerns. Other language features will be introduced as needed to support the remaining design activities. The basic syntax is presented in BNF notation with some regular expression shorthand to ease legibility. The shorthand symbols are:

{}: *Encloses comments.*

**:* *Any number of sequences of the preceding expression.*

+: *At least one or more sequences of a preceding expression.*

[]: *Any one of the set of characters enclosed within square brackets is applicable. A range of possibilities is indicated with "-": for example, [0-9a-z] indicates that any single digit and lowercase alphabet character can satisfy.*

Characters in bold typeface are reserved keywords in DSL. DSL adheres to Prolog's naming and syntax conventions.

The nonterminal symbols of DSL grammar are

```
integer:: = [0-9][0-9]*
real:: = integer.integer
number:: = integer|real|float
rel_op:: = <|>| = = | =<|> =
num_op:: = +|-|*|/|//|is
operator:: = rel_op|num_op||
literal:: = [a-z][a-z0-9]*
atom:: = literal|number
list:: = [member*]|[member|member]
member:: = <null>|dsl_name|variable|number
variable:: = [A-Z][A-Za-z0-9]*|[_][A-Za-z0-9]*
dsl_name:: = literal(parameter+)
             |literal
parameter:: = atom
             |variable
```

A DSL program consists of a combination of DSL constructs defined as

```
dsl_program:: = dsl_model_constructs+dsl_experiment_constructs*
```

The *dsl_model_constructs* constitute the applicable commands that are used for model description, and the *dsl_experiment_constructs* represent the commands used for experimentation and synthesis support. The former are defined below, whereas the latter are visited in the following chapters.

```
dsl_model_constructs::=module_definition
          |resource_definition
          |ho_module_definition
          |inheritance_definition
          |persistent_port_definition
          |path_declaration
```

The possible DSL modeling primitives are described in the following subsections.

Modules: DSL building blocks

The basic construct of DSL is an object entity termed a *module.* Modules are the primitive building blocks of the language (refer to Fig. 4.1 for the graphical equivalents). They form an implementation of the *model* object introduced earlier. A module possesses a name, a set of possible behaviors, and resources. They communicate with one another through messages, which are used as triggers within destination modules.

The basic DSL notation for a module is

```
module_definition::=module(dsl_name,[behavior*]).
```

where *dsl_name* is the unique identity of the module being defined, and is used extensively by DSL to resolve communication issues with-

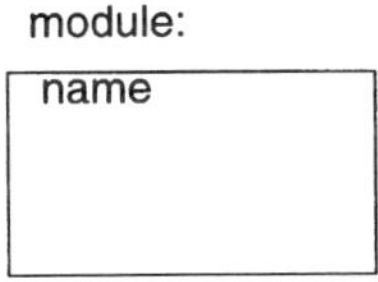

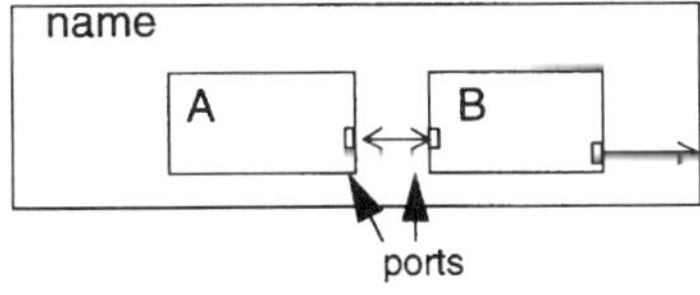

Figure 4.1 DSL constructs.

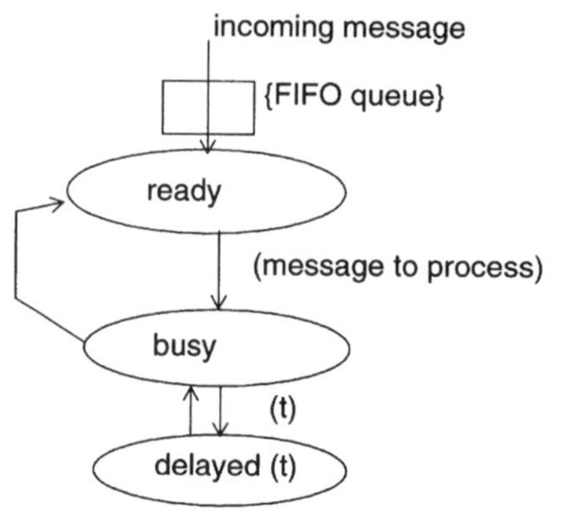

Figure 4.2 Basic states of modules.

in a model. The second argument, *behavior,* is a list of different behavior(s) the module may exhibit. The behavior(s) of the module represent the actions undertaken by the module when an event occurs. The occurrence of an event implies the arrival of a message from another module. The target module then attempts to execute the behavior that is associated with the message.

Modules operate within three conceptual states: *ready, busy,* and *delayed.* These are depicted in Fig. 4.2. The *ready* state indicates that the module can process a message. The *busy* state implies that it is currently processing a message, and a *delayed* state indicates that a module is suspended (delayed) and will commence processing the message after a time period *t.* When a module is in the *busy* or *delayed* state, arriving messages are queued within an input message queue, to be processed in turn by the module. Message execution within a module in the *busy* state is atomic. However, predefined interrupt messages (described later) can interrupt the execution process if the module is in a *delayed* state.

The description of behaviors will be deferred until a later section. Within a behavior, some actions may use or manipulate data structures local to a module to model states, local variables, etc. Such local data structures are represented within DSL through the *resource* statement. In accordance with the Prolog philosophy, local variables can also be lists, facilitating the management of structures such as arrays and queues. DSL commands to manipulate resources are defined in the section on data manipulation later in the chapter. A resource is (a predicate) of the form:

```
resource_definition::= resource(dsl_name, dsl_name, values).
values::= variable
    |atom
    |(values+)
    |list
```

For example, a statement such as **resource**(*module_name, res(P1,...,Pn), (V1,...,Vn)*) describes a resource local to a module (with name *module_name*). The resource (*res*) may be parameterized as above to define a set of *n* variables *Vi,* where each *Vi* can be any Prolog element, such as an integer, a list, or a string. The applicable operations that can be performed upon resources will be described in the behavioral modeling section.

Model composition

Composition is a basic requirement of an intermediate language. DSL permits the composition of modules into higher-order (HO) modules. Connections between modules are established through *ports.* A pair of ports constitutes a communication channel between the module and its environment. The DSL interpreter must deduce direction from the flow of information across ports. Ports are untyped, allowing different kinds of information to flow through the same port.

The graphical representation of a higher-order module (refer to Fig. 4.1) is similar to that of a module. The underlying subtlety is that higher-order modules encapsulate a set of modules and define interconnections between them. Hence using these constructs, more complex (higher-order) models can be composed from simpler ones. Higher-order modules possess no predefined behavior of their own; the interactions between the components establish the underlying behavior.

A DSL higher-order module is defined using the following syntax:

```
ho_module_definition::= ho_module(ho_name,module_list).
ho_name::= dsl_name
module_list::= [dsl_name*]
```

The *ho_name* is a *dsl_name* (Prolog structure), and the *module_list* is a list of modules composed within the higher-order module. Composition is not restricted to modules, but is also applicable to higher order modules. Hence members of a higher-order module may contain other higher order modules as well as modules; however, when referring to the constituent members of a given higher-order module, the term *module* will be used loosely to imply both. The interconnection of the constituent modules is achieved through a path declaration:

```
path_declaration::= path(module_x,module_y,[port_x,port_y]).
module_x::= dsl_name|variable
module_y::= dsl_name|variable
port_x::= dsl_name|variable
port_y::= dsl_name|variable
```

port_x and *port_y* are port names used to define a communication path between the two named modules. The connection implies a default direction from *port_x* to *port_y*; hence *module_x* by default serves as the source. Port names can also be written as variables, in which case no direction or specific port name is defined. In this case, the need for or creation of unspecified communication paths between two modules is delegated to the environment. The details of how the environment can manipulate the connections will be elaborated in the next chapter.

Port names can also be associated with local variables, which results in *persistent* ports. Such a port is maintained as a resource within a module under the reserved resource name *persistent_port:*

```
persistent_port_definition::= resource(persistent_port,
(port_name,values)).
port_name::= dsl_name
```

This capability permits ports to assume a value or state dictated by the most recent message that was sent from the port. Hence a state or value is said to persist on the port (similar to the behavior of a wire in hardware design). Persistent ports are suitable to model lower-level constructs such as wires and buses where the value of the entity (voltage levels or high impedance) may be of concern. This also permits different levels of representation (such as those found in hardware description languages) to coincide within a given model. Persistent ports impose a restriction on the connectivity. A module utilizing a *probe* statement cannot be connected to more than one persistent port per port. In Fig. 4.1, this implies that the port *pb* cannot have a connection to a persistent port other than *pa*. Since the *probe* statement uses only a single argument for the port name, more than a single connection would produce a nondeterministic result. However, there is no restriction on the number of connections *pa* may have.

Since paths are equivalent to Prolog predicates, unification and variable instantiation can provide a compact notation in defining some structures. For example, in the case of symmetrical connections, the *path* statement can be used concisely to define all module interconnections with a statement such as

```
path(proc(X), mem(X), [port(X), mem_port(X)]).
```

This statement will interconnect all modules named *proc(X)* and *mem(X)* through their respective ports (for example, *proc(1)* will connect to *mem(1),* etc.).

Relations can also be established through port definitions. For example,

```
path(proc(X), mem(X + 1), [port(X), mem_port(X + 1)])
```

indicates that *proc(1)* is connected to *mem(2)* through their respective ports. The *path* statements help define the necessary interconnections desired by the designer.

Module behavioral description

A module's list of behaviors describes all possible actions that may be undertaken by the module in response to a given message. The behavior of a higher-order module is a result of all possible interactions between the constituent modules. Module behavior is a procedural description of actions, which, within satisfaction of a set of constraints, may consist of:

- Communication initiation with other modules or internally
- Modification of resources or temporary variables associated with the module
- Timing-related directives

More precisely, the form is defined as

```
behavior::= (b_name:-condition*, action+)
b_name::= dsl_name
action::= communication*
      |data_manipulation*
      |timing
```

Under different conditions, behaviors can be viewed as a set of Prolog clauses, where each behavior can also be composed of multiple clauses. The head of a clause consists of the behavior name (*b_name*), a unary or *N*-position predicate. The body is a set of conditions and actions which are compound subgoals of the clause. It should also be noted that DSL conditions are Prolog subgoals to be satisfied, and the actions are subgoals that are always satisfied, resulting in some desired side effects (such as message generation and resource manipulation).

For a given behavior, a set of conditions can be evaluated before further execution of the behavior is attempted. This facilitates the description of *if-then-else* type constructs as well as supporting state-oriented behavioral description. The behavior selection may be based on built-in DSL predicates or Prolog conditional operators. Formally, conditions are described as

```
condition::= variable rel_op variable
     |number rel_op variable
```

```
|variable rel_op number
|number rel_op number
|check_res(dsl_name, value)
```

The *check_res* statement is used to test for a resource (*dsl_name*) or its value. For example, ***check_res**(counter,X)* will return the value (*X*) of a resource named *counter.* Alternatively, ***check_res**(Caller, telephone(5551211))* could return the name of a subscriber (*Caller*) associated with a specific telephone number. In either case, if there is no match, the predicate will fail and another satisfaction of the behavior will be attempted.

DSL uses the Prolog programming style for resolving ambiguities and testing conditions. Hence, both declarative and procedural styles are possible in specifying behaviors. A behavior with the name *b(X),* for example, is an unconditional attempt to satisfy the corresponding behavior statements. However, code in the form

```
b(1):-c1, a1.
b(wait):-c2, a2.
b(X):-c3, a3.
```

gives multiple behavior names that DSL will attempt to satisfy for the arguments "*1*", "*wait,*" or otherwise. In addition, a set of conditions *c1, c2,* and *c3* determines whether the action (*a1, a2, a3*) for the given clause will be attempted (since actions are always true, the conditions act as guards against erroneously selecting actions).

Now that the basic behavioral styles have been introduced, the possible actions available within a behavioral description are described.

Communication primitives. Communication is the main means for changing state within DSL models. The main action that invokes communication is the *send* statement. A DSL interpreter will resolve the destination and schedule the message associated with the statement within the destination's input queue. The format of the *send* statement is as follows:

```
communication::= send(destination, port, message)
      |send(message)
      |execute(message)
destination::= dsl_name|variable|list/[*]
port::= dsl_name|variable
message::= dsl_name
```

The first form of the *send* statement has three possible parameters: *destination, port,* and *message.* The *destination* is the name of a destination module. This may be written as a list to specify multiple destinations. For example, if the destination is *[a,b],* the module will send the same message to modules *a* and *b.* An empty list *[]* will result in a message being sent to the first module capable of interpreting the message. The destination can also be specified as *[*]*, which has the potential effect of sending a message to all modules capable of processing the message.

The *port* is the name of a communication port of the source module from which the communication is to be initiated. The *message* is the name of the message to be sent. This can be any valid behavior name that is comprehensible by the receiving module.

The second form of *send* is for internal messages, and is essentially a short version of the first, where the destination is the source module. Using this form of the statement is the usual way of generating a continuous stream of messages with possibly varying parameters. For example:

```
(signal(AMPLITUDE):
    delay(FREQUENCY),
    send(_,output_channel, waveform(AMPLITUDE)),
    send(AMPLITUDE-1))
```

This segment of code will generate a message named *waveform* with a parameter (or attribute) *AMPLITUDE* at a minimum time period defined by a variable named *FREQUENCY*. The value of *AMPLITUDE* will be diminished by 1 for each new message. The last line of the code ensures that a continuous stream of messages is sent.

The first two fields for the first form of *send* are optional, which provides extreme flexibility in the way communication can be initiated. The possible actions are described by the following cases (parameters are undefined unless they begin with a lowercase letter):

Case 1. All parameters are given: ***send****(dest, port, message).*

Action. *message* will be sent from the source module to the module with a name *dest,* through the source's port called *port.* A special case is when the destination is *[*].* In this situation, the message will be sent to all modules that are connected to the source's port. This is equivalent to a broadcast message in some systems.

Case 2. No destination name is given: ***send****(_, port, message).*

Action. *message* will be sent to the first module that is capable of executing it and is connected directly or indirectly to the source module's *port.*

Case 3. No port name is given: ***send****(dest,_, message).*

Action: This is essentially the same as case 1. Ports are not necessary for communication, only for synthesis later in the process. This type of *send* statement will generate pseudo-communication channels within the DSL environment which can be used by a synthesizer to construct data paths.

Case 4. No port or destination given: ***send****(_, _, message).*

Action. This is similar to case 2. However, any module that can interpret the message will be selected. This mode should be used with caution, since the communication paths are determined solely by the environment and not by the designer. Subsequently, connections between existing modules can be inadvertently established.

A message generated through a *send* statement will always be placed in the input (message) queue of the destination module. Hence there is an implicit causality associated with message communication: Each message will be processed in the order in which it was received.

The other communication primitive available is the *execute* statement. This statement is similar to an internal *send* statement; however, it bypasses the input message queue of the module. Consequently, the message will be processed immediately.

A combination of *send* and *execute* primitives can implicitly establish different message orderings. Figure 4.3 presents some possible message combinations. The figure assumes that two modules (*a* and *b*) are

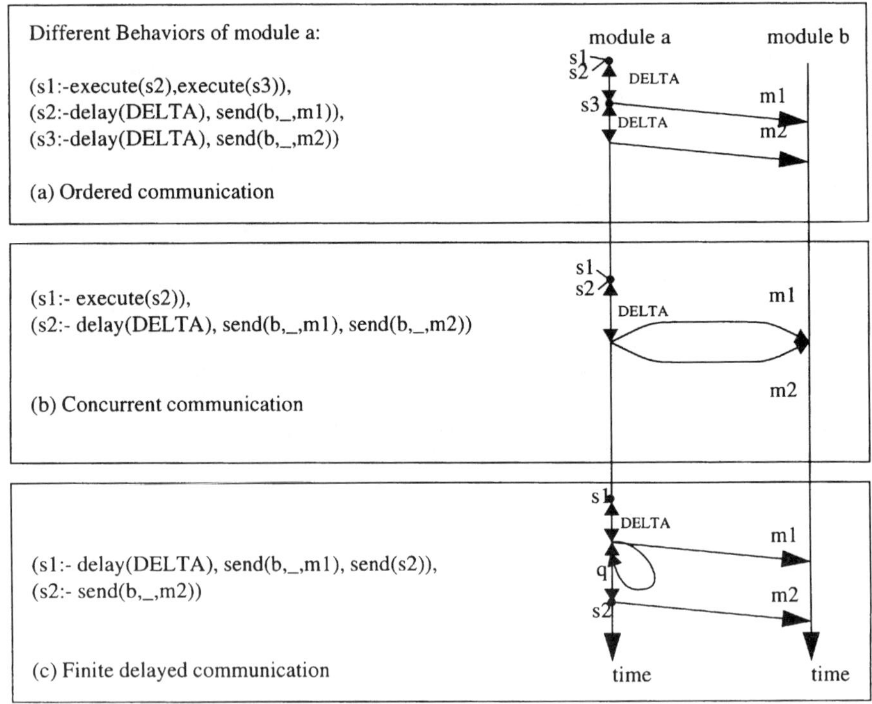

Figure 4.3 Possible communication between modules.

communicating. In particular, module *a* is actively sending two types of messages, and the diagram indicates the relative times at which the messages arrive at module *b*'s input queue.

The first case shown is the effect of multiple *execute* and *send* statements. The action is that of an ordered transmission of messages at predefined intervals. Three sets of behavior, each with a discrete delay of *delta,* are assumed. The first behavior (*s1*) generates two *execute* statements for behaviors *s2* and *s3*. The latter two behaviors use *send* statements to generate messages *m1* and *m2,* respectively. The destination module will receive the two messages *delta* time units apart in the order in which they are sent. It should also be recalled that receipt in the input queue does not imply that the messages will be immediately processed. This is dependent on the internal state of the destination as well as the number of messages waiting in the input queue.

The second case depicts a concurrent transmission of messages *m1* and *m2*. This is achieved through behavior *s2*. Although message *m1* is listed before *m2* within the behavior, the order of arrival at the destination is nondeterministic. The DSL interpreter will schedule the two *send* statements at the same time in any order. This is due to the fact that apart from the *delay* statement, the execution of DSL statements does not consume any simulated or modeled time. Hence, for all intents and purposes, the two *send* statements occur at the same simulated time.

The final case shows the use of a combination of *send* statements. Behavior *s1* sends message *m1* after a delay *delta,* then sends an internal message (*s2*). The latter will be queued and processed after a finite queuing time *q,* and *s2* will generate message *m2*. The effect of the internal *send* is to schedule *m2* after *m1,* but with an unknown delay *q* between the two.

Data manipulation. Data manipulation actions can change the values of internal variables defined by resources or temporary variables local to a module. The actions can utilize any valid arithmetic and list operators of Prolog. In addition, data values used within resources can also be manipulated within DSL statements.

The following comprises a description of all data actions.

```
data_manipulation::= set_definition
     |create_definition
     |remove_definition
     |probe_definition
     |arithmetic
     |built-ins
set_definition::= set_res(dsl_name, values)
create_definition::= create(resource_definition)
remove_definition::= remove(resource_definition)
```

```
probe_definition::=probe(dsl_name, variable)
arithmetic::=variable operator expression
expression::=number
    | (expression operator expression)
```

The first four data manipulation statements are those related to resources. the *set_res* statement permits the assignment of a value to the respective resource. For example, ***set_res****(state, busy)* will set the resource *state* to the value *busy.* The ***create*** action will instantiate a new resource. Hence, ***create****(buffer(1), empty)* will define a new resource named *buffer(1)* with contents *empty.* The ***remove*** action will remove (or disassociate) a resource from a module. For example, ***remove****(buffer(1),full)* will remove the named resource, whose contents are defined to be *full.* It should be noted that this operation has the effect of removing a defined fact from the underlying Prolog database. Hence, if there are multiple entries with the same resource name, only the oldest copy of the resource will be removed.

The *probe* action is reserved for resources defined as persistent ports. The invocation of the action results in the variable's being bound to a local value maintained for the persistent port. The statement permits a module connected to another module's persistent port to obtain the state of the port. The nuances of this representation are illustrated in Fig. 4.4. In the example, module *A* has a persistent port *pa,* through which it is connected to another module *B.* The implication is that there is a resource associated with module *A* that contains a state variable *V.* At some point in module *B*'s behavior, a ***probe*** statement is encountered. The effect of this is to access the value of *V* from the persistent port and bind it to the local variable *ST.* In the figure, this is implied through the connection that exists between ports *pb* and *pa.* Persistent ports can be read only by other modules.

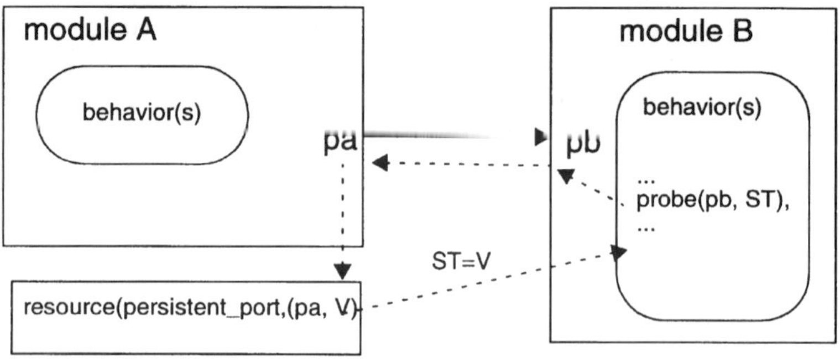

Figure 4.4 Example of persistent ports.

Changing the value can be performed only by the module associated with the port (in this case, module *A*).

The *probe* statement is useful for depicting lower-level interactions. For example, consider a model of a processor connected to a bus and a memory unit. Assume that detailed operation of the memory is of interest and that the processor model generates messages representing a state of the bus (for example, *read_memory* could imply a "0010" bit level on the control bus). If the bus were to be represented by a module with a persistent port, the processor could affect the state of the port with its messages to the bus, and the memory module could utilize *probe* statements to latch onto the required values.

There are additional behavior predicates in DSL to help with the manipulation of data and creation of standard data structures such as queues. These are summarized below:

```
built_ins::=
     not(dsl_name)
    |member(literal, list)
    |remove_element(literal, list, new_list)
    |list_length(list, variable)
    |append(atom, list, new_list)
    |iqsort(list, new_list)
    |last_list(list, atom)
```

The statements ***not*** and ***append*** correspond to Prolog predicates. The first returns the negation of the argument, and the latter appends an element to a list. The ***member*** statement is true if its argument *literal* is a member of the list *list*. The statement ***remove_element*** removes the first occurrence of an element *literal* within the list *list* and returns the sublist *new_list*. ***list_length*** determines the number of elements (*variable*) that are within the list *list*. **iqsort** is a quicksort algorithm for the sorting of the elements of *list*. Elements are sorted in ascending order and placed in *new_list*. Finally, ***last_list*** identifies the last element *atom* from a queue named *list*.

Timing. Up until now, timing has not been elaborated upon. Timing behavior in DSL is limited to simulation; hence there is no extensive formal timing model as part of the language semantics. However, delays imposed on module behaviors can be simulated with the use of the *delay* action, as would be expected in a simulation-supported paradigm. A *delay* action suspends the module for a given period of simulated time. After the elapsed time, the next action in the behavior is executed. The syntax of this action is

```
timing::=delay(number)
    |delay(variable)
    |c_function, delay(variable)
c_function::=call_c(atom, variable)
```

The action can utilize a numeric value explicitly or implicitly (as an unbound variable). The former simply implies the use of a numeric parameter such as *delay(5.4),* which will suspend further operation within a module for 5.4 time units. The latter is a case in which a variable may be dependent upon an externally passed parameter—for example,

```
(Behavior_name(T1):-delay(T1),.).
```

In this case, *T1* is passed to the module as a delay parameter. The *call_c* action can also be used in conjunction with a delay action to access random variables from a C library. Two examples of how random messages may be generated are given in Fig. 4.5. The first uses a list (*RANDOM* | *REST*) to select a random number (a table lookup scheme could also have been used), suspend operation for that time period, generate a message (*stimuli*), and then access the next number. This structure has the advantage of replicating the same sequence of random numbers for every simulation scenario. The second example is that of a random-number generation. A system call to C is used to access a random-number generator (*rand*) and obtain a number between 0 and 1 (which is scaled afterwards). A Prolog predicate *call_c(Function,X)* is used to pass the random value from C to Prolog (some implementations of Prolog have predefined random-value predicates that can be used directly). The value is then used to delay and generate a message as before. The difference here is that each simulation scenario will utilize different random values, and the timing delays will not be identical.

Timing issues for modules are resolved through scheduling during simulation. The execution of a communication statement within a module's behavior is handled directly through the simulator after the

```
module(random, [
(start([]):-
        print('Finished random messages')),
(start([RANDOM\REST]):-
        delay(RANDOM),
        send(_,_,stimuli),
        send(start(REST)))
]).
```

(a) Predetermined random numbers

```
module(random, [
(start:-
        call_c(rand, RANDOM),
        delay(RANDOM),
        send(_,_,stimuli),
        send(start))
]).
```

(b) Utilizing function calls

Figure 4.5 Example of random-number generation.

destination name has been resolved (Chap. 5 describes the simulator data structures in detail). Similarly, a *delay* statement will cause a rescheduling of the current message being processed, until a specified time period. During this period, no further messages may be processed (i.e., the module is in a delayed state). There is, however, one exception to this rule. A default class of messages are defined within DSL to act as interrupt messages. These are defined through the statement

interrupt_declaration:: = **isa(***message*,**dsl_interrupt).**

In this declaration, *message* is defined as a priority message. If a module is in a delayed state, such a message will cause the simulator to immediately remove the current message and schedule it after the priority message. After completion of the priority message, the interrupted message will once again delay the module. In the busy state, messages cannot be interrupted. A priority message can be interrupted by another priority message (there is no notion of levels of priority).

Inheritance and hierarchy

The next property the language must possess is support for inheritance and hierarchical-model construction. DSL permits modules to inherit behavior (and structure) from other modules (and higher-order modules)—this expedites model reuse. A module that is written to be sufficiently generic and that can be reused by different parents is termed a *generic* module. By convention, such module names are preceded with "*generic_*" to stress this property.

Creation of generic modules is simplified by relaxed coupling between modules. A module does not necessarily require any knowledge about its destination; hence communication can be defined without destination names. Also, since ports permit different assortments of messages to pass through them, interconnection is also simplified. Model coupling is achieved by higher-order modules accessing the generic components through the library. Generic modules will be explained in detail in the library support subsection in Chap. 5; however, it should be noted that if an inheritance mechanism exists and if the superclass (generic) module does not use communication primitives that embody destination names, then many different subclasses of modules may reuse the superclass definition. Expanding upon this notion, different reuse capabilities can be observed, as shown in Fig. 4.6.

The figure shows three different cases of inheritance. In each case, module A is the superclass module and B the subclass. In the first case, the communication primitives do not specify destination or port

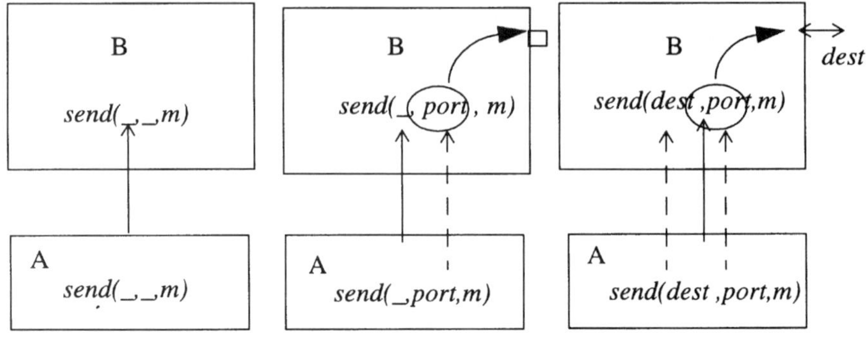

Figure 4.6 Module inheritance and reuse.

names. Hence the definition of module *A* resembles a truly generic module, whereby all the behavior is inherited as is.

The second case assumes the use of port names (but still no destination names) within the *send* statements. This is still a useful generic module within DSL. The environment has the ability to adapt the subclass to the additional demands imposed by the superclass. In this case, new port definitions may be required; hence additional port connections can be generated by the design environment.

The final case is not useful for generic module design. The communication primitives refer to destination names, and hence there is an added assumption that the subclass must be able to communicate with predefined modules. If, however, the communication is restricted to the boundaries of the superclass (within a higher-order module), then there is no problem in inheriting.

The mechanism for inheritance is defined through the *isa* statement as follows:

```
inheritance_definition:: = isa(module_x,module_y).
```

The statement is used to establish that entity *module_x* is a subclass of entity *module_y*. *module_x* can also be defined to possess additional behavior local to its particular function. The statement also permits multiple relationships to be defined in one statement. For example, ***isa***(*processor(X), cpu(risc)*) identifies that all modules defined with the name *processor(X)* (where *X* is a variable) inherit properties associated with a superclass module called *cpu(risc)*. Module inheritance is not limited to a single source; hence multiple inheritance is allowed, as in the following:

```
isa(processor(xyz), cpu(risc)).
isa(processor(xyz), memory(32M)).
```

In this example, *processor(X)* is defined as before; however, a module *processor(xyz)* is also indicated as inheriting some additional properties from another module (*memory*).

The ***isa*** statement can also be used to identify relationships between messages. For example, the following two lines define relationships between three different messages:

```
isa(update(ADDRESS,VALUE), msg_reqst(SITE, component(ADDRESS),
update(VALUE))).
isa(msg_reqst(SITE,component(ADDRESS),update(VALUE)), memory_write
(SITE + ADDRESS,VALUE)).
```

The first identifies a relationship between *update* and *msg_reqst*. This may represent a correspondence between a high- and a low-level protocol where an additional parameter (*SITE*) is provided by the respective module to interpret the message. In this case, the higher-level update message will be executed as a *msg_reqst* message at the destination module. There is a further relationship, identified by the second line above: Message *msg_reqst* can also be broken down to another message *memory_write,* which may be a message executable by a memory module. In this manner, a high-level representation of a message can be sifted through different levels of abstraction.

The inheritance scheme for messages is adequate for messages that can be mapped to one another. However, if a message requires more complex interactions at a lower level, such as multiple messages, acknowledgments, and error-checking conventions, it is best to use a module that specifically models the desired protocol.

Module behavior as predicate/transition nets

The semantics of most elements of a DSL module can be described in terms of a predicate/transition net (PrTN) (Genrich 1981). The implication of this is that some properties of modules can be formally examined using analysis methods available to PrTNs. The analysis possibilities are described in Chap. 5, "Design Execution," which can provide for verification of modules before placing them in a module library. This section will present the PrTN model of a module and its limitations.

Petri-nets have been used in many areas to analyze hardware and software systems. The formalism provides for a relatively powerful way of defining asynchronous concurrent communication. One of the main complaints voiced about Petri-nets has been the problem of state-space explosion and the complexity of the generated graphs. These concerns are minimized within DASE with the use of predicate nets to reduce net sizes and also by observing some restrictions to the general Petri-net formalism.

- The first restriction is that modules exhibit a safeness property akin to that of Petri-nets. This is observed from the fact that a module processes one behavior at a time, which is equivalent to a known bounded maximum number of tokens in a Petri-net or a strict PrTN.
- The number of possible types of messages a module can accept or generate is finite and can be parsed. This implies that a finite number of input and output places for each module can be constructed. If DSL messages are represented as PrTN tokens, this also permits the definition of finite token sets.

A DSL module can be represented by a PrTN as shown in general form in Fig. 4.7. The diagram gives the basic structure of the net that can be generated for a given module. At the top, an input place (shaded in black) where tokens annotated with possible message names arrive is defined. The assumption is that only one token may reside in the input place at a particular time. To guarantee that no new message is processed by the net, a busy place is explicitly indicated where a token can be grabbed by active message tokens and released upon completion of their behavior.

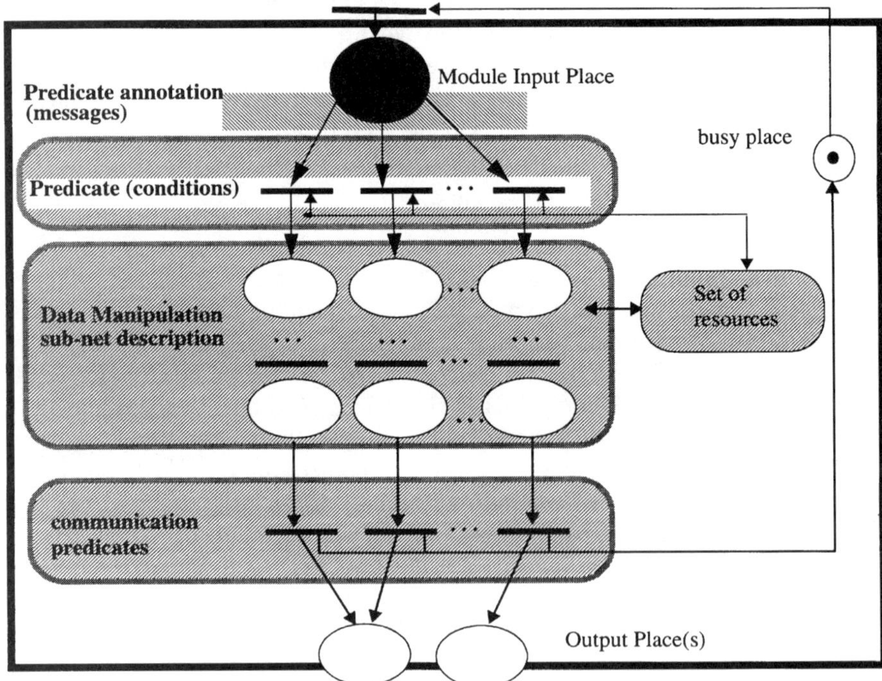

Figure 4.7 Predicate net representation of modules (1995 IEEE).

The initial arcs emanating from the input place are annotated with different possible message predicates; hence each token type is directed to a different subnet corresponding to the particular DSL behavior. Any conditions demanded by the DSL model for a given behavior are annotated upon the first set of transitions. In the event that a condition is required from a resource, a bidirectional annotated arc is created to a place maintaining a token for the given resource (within the set of resources block). The output arcs of these transitions connect to a data manipulation subnet which models the data manipulation statements of DSL. Eventually, arcs connect from this subnet to a set of transitions that make assignments to tokens as necessary, modeling a DSL communication statement, to generate a token to an output place.

The individual DSL behavior statements can be translated into PrTN net elements as depicted in Fig. 4.8. As can be seen, each of the basic DSL elements can be represented by a subnet. Each subnet can connect with others through intermediate places between them. The connections to the intermediate places are made by input and output arcs within each subnet. These arcs are shown as partly connected to each transition in the subnets of Fig. 4.8.

There are two limitations of this kind of translation. The first is that *delay* statements are ignored, and hence timing is not explicitly captured in the net representation. Another limitation is that the name of the resource must be given in the *create* statement. In DSL it is possible to define a behavior such as

```
b(Resource, Value):-create(Resource, Value).
```

Since this resource name depends upon an external type (supplied by another module), a corresponding place cannot be identified within the PrTN. A similar problem also exists with the *probe* statement, since it requires a value from an external module.

Example. Observing the limitations, an algorithm for translation from DSL behavior to a PrTN is given in Fig. 4.9. A small example is presented below to demonstrate the algorithm and clarify the use of the annotations. Larger nets are presented in the case studies described in later chapters.

Assume that a DSL module *timer(Value)* is to be translated to a PrTN, where the module is described as

```
module(timer(Value),
[(clck:-check_res(count, M),
     Value>M,
     New is M + 1,
     set_res(count, New)),
(clck:-send(Dest, out, alarm)),
     execute(reset)),
(reset:-set_res(count, 0))]).
```

DSL Statement: **PrTN Representation:**

Condition statements

check_res(Resource($r_1,\ldots,r_z$), ($v_1,\ldots,v_y$))
i.e. check_res(count(1), Value).

$V_n\ \Omega\ V_m$
where Ω is a valid Prolog operator.
i.e. N < New.

Data Manipulation statements:

set_res(Resource($r_1,\ldots,r_z$), ($v_1,\ldots,v_y$))
i.e. set_res(count(1), 0).

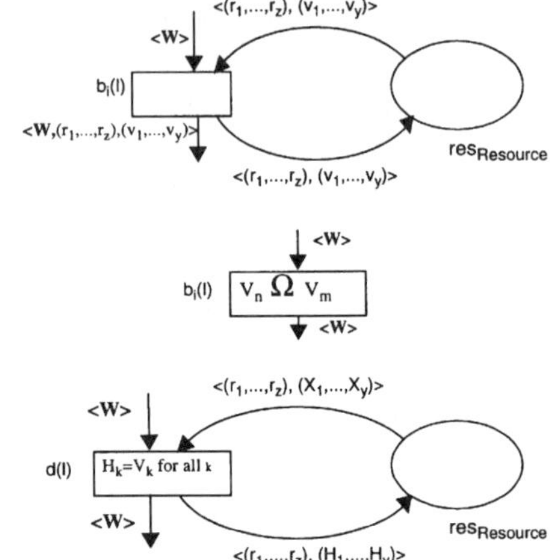

create(Resource($r_1,\ldots,r_z$), ($v_1,\ldots,v_y$))
i.e. create(buffer(8), (1,0,0,1)).

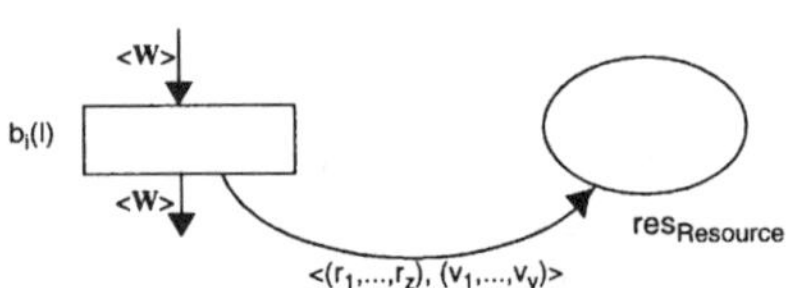

remove(Resource($r_1,\ldots,r_z$), ($v_1,\ldots,v_y$))
i.e. remove(pointer(1), VAL).

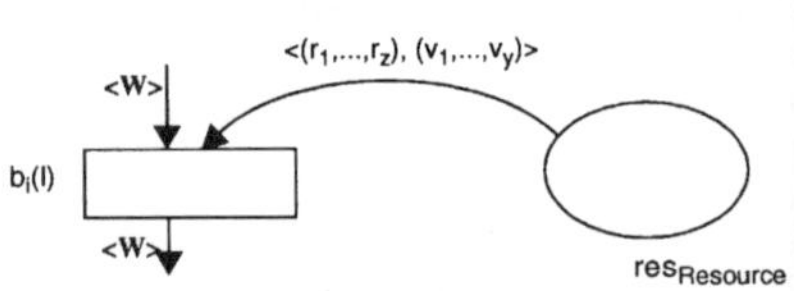

Communication statement:

send(Dest($d_1,\ldots d_n$), Port($o_1,\ldots,o_m$), Message($a_1,\ldots,a_z$)).

note:
(execute(M) is simply an arc back to the input place)

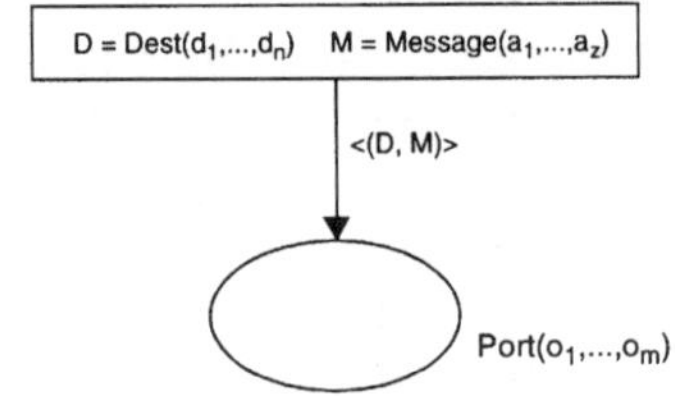

Figure 4.8 Predicate net correspondence with DSL statements.

The DSL behavior describes a timer that counts the number of *clck* messages it receives and transmits an alarm message (through its *out* port) when the count has reached a limit *Value*. At this point, the count is reset to 0 and the process repeats. An external reset message will also reset the count.

```
Define a set T = {} and P={};
For a given module name with arguments e_1,...,e_n
  create an input place labeled in_name and add it to P;

For each behavior b_i(a_1,...a_k) do
{ create a transition node labeled as b_i(l) where 0 < l < n such that b_i(l) ∉ T , and add it to T;
  current_transition = b_i(l);
  current_label = (a_1,...a_k);
  create an arc with annotation <b_i(a_1,...a_k)> from in_name to b_i(l);
  For each DSL statement of the clause defined by behavior do
  {
  If a condition statement is encountered then
        if the condition = check_res(Resource(r_1,...,r_z), (v_1,..,v_y)) then
        {         if a place res_Resource does not exist create it;
                  create and annotate arcs as in figure 4-8;
        else
        {         annotate the transition b_i(l) with the condition}
  if a data manipulation statement is encountered then
        {create transition ti, place pi and arc(s) and annotate as required by figure 4-8;
                  connect arc between current_transition to newly created place pi;
                  annotate arc with current_label;
                  current_transition = ti;
                  current_label = {current_label appended with annotation of input arc to ti.}
        }
  if a communication statement is encountered then
        {create transition si and annotate as required by figure 4-8;
        create arc to appropriate output place;
        }
  }
}
```

Figure 4.9 DSL-to-predicate-net translation algorithm.

The behavior can be represented as a PrTN. The translation algorithm generates the PrTN net shown in Fig. 4.10. The figure omits the busy place, which is part of a standard module's behavior. The input place accepts the incoming messages, and one of three transitions *(reset(1), clck(1), clk(2))* is fired depending upon the message name. A place is required to hold a token relating a value to a resource (labeled res_{count}) "count." The two possible conditions for the behaviors for *clk* are represented by the *clk(1)* and *clk(2)* transitions. In the event of a *clk* token in the input place, *clk(1)* will first be tested, and then, if needed, *clk(2)*. The external message generated by the net is a result of following the *clk(2)* path, where an alarm message is sent to *Dest.*

The graph can be depicted with PrTN tools (such as the PROD language developed at the Helsinki University of Technology; see Gronberg 1993) for further analysis.

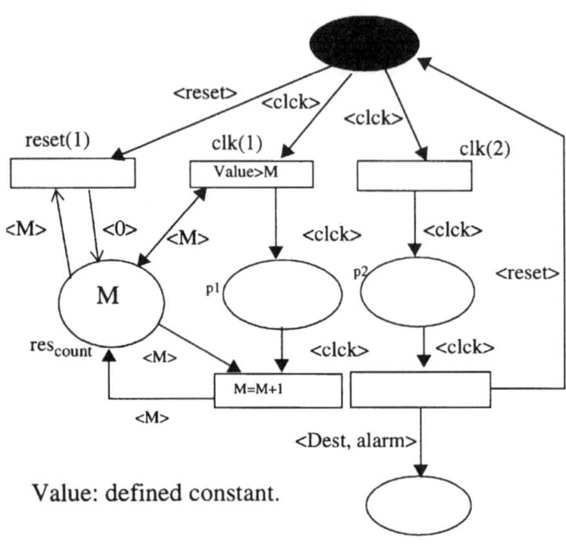

Figure 4.10 Timer example.

DSL Modeling Example

This is a good point to revisit the DTSS example which was introduced in the previous chapter. At the architectural level of design, typical design considerations include traffic, the number of lines required for the switch, and verification that a call will be set up. For the designer to explore different trade-offs in a rapid and intuitive manner, key reusable or generic components must be available. For example, the number of input lines is an important factor in the switch size. The design criterion is likely to affect the number of interface units to use and the size of the switch element. Hence such a design parameter must be accessible to the designer. Whether access is provided through a library support system (as is recommended in the next chapter) or through careful coding of the behavior parameters is an environment design issue.

To address high-level design concerns, a higher-order module description of the digital telephone system (*dts*) switch is presented. The module should possess all the necessary parameters which the designer may want to manipulate at this point. This can be accomplished using DSL as

```
ho_module(dts_switch(NUMBER, INPUTS, CARDS, CHANNELS),
   [switch_element(NUMBER), interface_card(Y)]).
```

The parameters for *dts_switch* are relevant to the system designer. These are

- *NUMBER*. This is an identifier for the switch. It is used to identify and configure different switches within a network of interconnected ones.
- *INPUTS*. This defines the number of input (and consequently output) lines that a switch should have.
- *CARDS*. This defines the number of interface units the designer may require.
- *CHANNELS*. This defines the number of channels the serial lines should comply with (the default is 24, which is DS1).

The description of the structural details for the two main hardware components of the switch is presented below, with some elaboration of the behavior.

The description of the software element (telephone services) is given as an example of a behavioral specification in DSL. The complete design description in DSL can be found in the appendices.

The generic switch element

The design presented in this section is one that is applicable to many large DTSS designs. The example concentrates on the call-setup operations, and so other operations that are important in the switch design (such as customer billing, error processing, maintenance, and fault tolerance) are not included in the example. However, these are easily added as inherited (software) modules in a similar manner to the telephone services described later in the section.

The real-time design concerns for this element can be numerous and complex, depending upon the size of the switch. The objective of the generic switch element is to capture the serial streams of data from input ports, identify the incoming time slot and port, determine the destination port and time slot, and ensure that the right incoming data get transferred to the correct output stream. Since the number of input and output ports is variable, it is useful to separate their functionality into separate modules. The design will also require a control processor(s) to manage all the routing and call-setup functions. To avoid congestion on the control processor bus, a separate bus to move data from the input to the output is desirable.

Addressing these design issues, a typical *generic* switch element model is given in Fig. 4.11. The *generic* prefix is used for models that are designed with the intent that, if a library support system were available, the model could be reused within other designs.

The modules are described below:

- *line_in:* This module is the interface for the incoming message stream. Incoming serial channels are captured as byte-long words

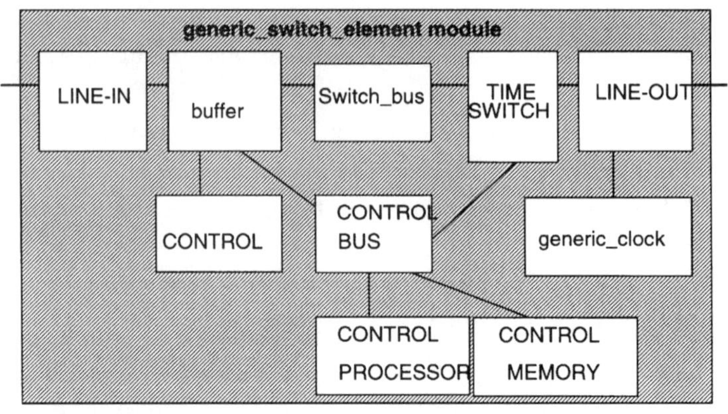

Figure 4.11 Generic switch model.

(or voice samples) and sent to the *buffer* module for temporary storage.

- *buffer:* This is a temporary storage facility for all the *line-in* modules, which implies that the buffer size is dependent upon the number of *line-in* modules.
- *control:* This module provides the timing control for the *switch_bus* to transfer data from the input buffers to the output. Control is also provided for use by the *control_processor.*
- *switch_bus:* This is a model of the bus between the input buffers and the output (*time_switch*).
- *time_switch:* This module, under *control_processor* control, latches onto data from the *switch_bus* and transfers it to the appropriate *line-out* module.
- *line_out:* This module depicts the output interface. The module contains a buffer for each channel it supports (e.g., for DS1 there are 24 buffers). New data from the *time_switch* overwrite relevant buffers, and information from each buffer is transmitted in sequence under control of the *generic_clock* module.
- *generic_clock:* This module provides the timing requirements for DS1-type transmission equipment. The behavior of the module will be further elaborated on in this section.
- *control_processor* and *control_memory:* The module routing and call-setup control is maintained by these modules. The memory contains the routing tables for call connections as well as the control software.

```
1.0 module(generic_clock(INDEX, NO_OF_CHANNEL), [
1.1         (clock_count(NO_OF_CHANNEL):- send(clock_count(0)),
1.2                 send(_,clock_port,frame_cycle)),
1.3         (clock_count(NEW):- NEW<NO_OF_CHANNEL,
1.4                 Clock_rate is (125/NO_OF_CHANNEL),
1.5                 delay(Clock_rate),
1.6                 COUNT is NEW+1,
1.7                 send(_,clock_port,clock(NEW)),
1.8                 send(clock_count(COUNT))) ]).
```

Figure 4.12 Generic clock code.

The behavior. As an example of DSL behavior, the details of the *generic_clock* module are given in Fig. 4.12. This segment of code is typical of library modules. The first segment (lines 1.0 to 1.8) is the behavioral description of the module. The module processes only one message: *clock_count(NEW),* where *NEW* is an undefined parameter or equals *NO_OF_CHANNEL* (a parameter passed to the module). The parameters of the module are given in line 1.0. *INDEX* is an identification number given to the clock module, and *NO_OF_CHANNEL* is the number of channels the clock is to support. The latter permits the module to produce timing control for a wide variety of transmission components.

Lines 1.3 to 1.8 demonstrate the typical actions that occur upon receipt of a *clock_count(NEW)* message (where *NEW* = <*NO_OF_CHANNEL*). Line 1.4 establishes a *Clock_rate* for each channel, and in line 1.5 the clock module is delayed for this period of time. After the delay, lines 1.6 and 1.7 increment the value of *NEW* and send a *clock(NEW)* message to any module connected to its *clock_port.* In line 1.8 an internal *clock_count* message is transmitted with the incremented value of *NEW* to start the cycle over again.

Lines 1.1 and 1.2 guarantee that the module provides a modulo-*NO_OF_CHANNEL*-based timing. If the *clock_count* parameter is at the limit (*NO_OF_CHANNEL*), an internal message *clock_count(0)* resets the parameter. A *frame_cycle* message is also generated to modules connected to the clock's *clock_port.* This message advises the modules of the start of a new frame (or a sync bit in DS1).

The second segment of code (lines 1.9 to 1.13) defines starting conditions for this module. The required start message in this case is a *clock_count* message with an integer value, used to start the clock module. This value can be anything in the range 0 to *NO_OF_CHANNEL*. If a value is not provided by the user as in line 1.13, the environment obtains the value from the instantiation of the module (lines 1.10 to 1.12) if possible; otherwise an error message is given to the user. In this case, the instantiation of the generic module would provide the *NO_OF_CHANNEL* value which is a valid parameter for the *clock_count* message. Details of how this can be accomplished will be deferred to the next chapter.

The interface unit(s)

The design demands for the interface units are less severe. These units must provide the basic services that an individual telephone will require: services such as dial tone, ring-back tone, and capture the dialed digits. These services can be provided in many ways. In this example, they will be provided by a programmable *controller* module. The module's "program," and hence its behavior, is defined by another library module: *sub_server* (for subscriber server). Direct interactions with the telephone can be altered by changes to the latter module.

The interface units also require a *line_interface* module which will provide the necessary transmission capabilities between the *switch_element*. The behavior of this module is identical to that of the combined *line_in* and *line_out* modules within the *generic_switch_element*. Also, a timer module is defined (an instantiation of the *generic_clock* module) to provide the appropriate timing control for the transmission components. The DSL representation of the interface unit is provided in Fig. 4.13. The *sub_server* module describes the con-

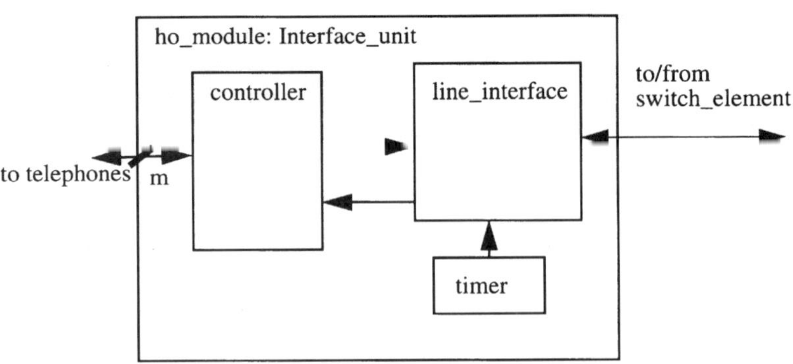

Figure 4.13 Interface unit model.

trol and management of the telephones connected to the interface unit. Its behavior will be shown during the description of the telephone call-setup services in the following section.

Software constructs—telephone services

The example assumes that the telephone call-setup services provided by the switch are controlled through the *control_processor* modules in the *generic_switch_element.* This is a design choice. Alternatively, most of the services could be off-loaded to the *interface_units* or to another processor. These alternative scenarios can easily be incorporated within this example by inheritance of different parts of telephone services through other modules. In addition to basic call-setup services, the switch must satisfy the basic requirements from the subscribers (telephones), as was described in Table 3.1. Table 4.1 depicts a corresponding set of DSL messages associated with each of these requirements.

The breakdown of the requirements into high-level DSL constructs is intuitive. This is because the DSL model description is at a level of abstraction comparable to the requirements (as should be the case to support architectural design). The DSL messages will be used in the communication of the software modules *pots_server* and *sub_server.* The two provide the call-setup and telephone control functionality, respectively.

The *sub_server* module views each telephone it services as being in one of four states, as shown in Fig. 4.14. When a telephone is not

TABLE 4.1 DSL Correspondence of Service Requirements

No.	Description of service	DSL behavior name
1.	-Detect an incoming call (off-hook) from source agent -Provide dial tone	*phone_off_hook(A)* dial tone
2.	-Collect digits from source agent	*dial_dest(A,B)*
3.	-Translate digits	*call_dialed(A,B)*
4.	-Select terminating agent	*status_response (B,STATE)*
5.	-Establish connection between the two agents	*called_state(A,B)*
6.	-Send ring tone to terminating agent -Send ring tone to originating agent -Send busy tone to originating agent	ring tone ring tone busy tone
7.	-Detect answer (off-hook)	*responded_call(B)*
8.	-Detect a disconnect (on-hook)	*phone_on_hook(B)*

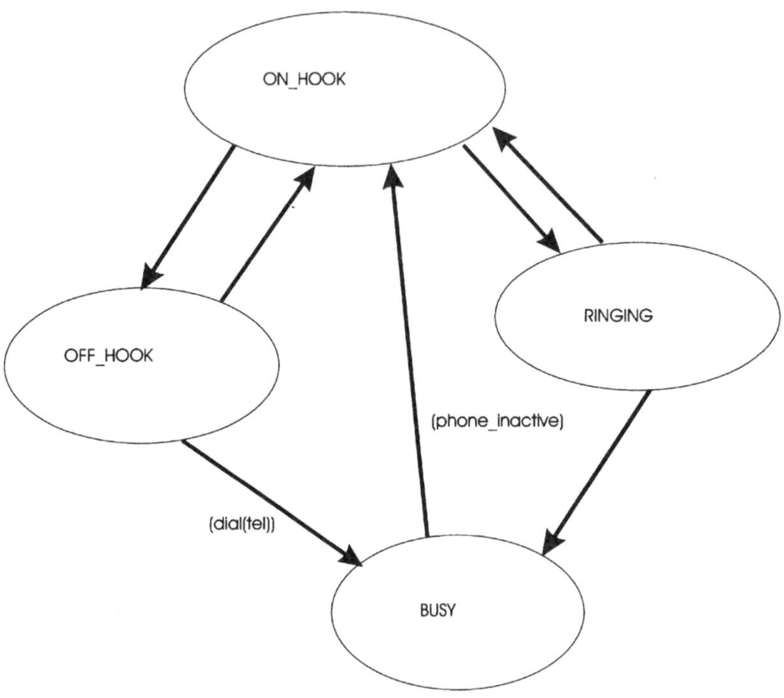

Figure 4.14 Telephone states managed by the *sub_server.*

being used, it is termed to be in the *on_hook* state. When the handset is lifted, the switch detects this, and the telephone is said to be in an *off_hook* state. After a number is dialed, the telephone is in the *busy* state, where it remains until the user hangs up. Finally a *ringing* state indicates that the telephone is being called. These states are affected by interactions between the *sub_server* and *pots_server* modules and the telephones.

Telephone services may require communication between elements of the switch and other switches. For example a call originating on one switch could be trying to ring a party connected to another switch. To transmit this type of signaling information, the switches will reserve channel 0 (the first channel in a frame) as the default communication channel. Hence all telephone service messages passed from the *generic_switch_element* to the *interface_units* will be through this channel. Alternatively, a separate line can be allocated specifically for signaling information—which is another design choice.

The call-setup service routines defined within a *pots_server* module will eventually be synthesized to software, and hence will not affect the DSL model in a structural manner, its behavior being inherited by the *control_processor* within the switch. It will however, interact with the *sub_server* module to manage the connections. A clearer understanding of the interrelations between the two is presented in the Fig. 4.15.

The *sub_server* functions are shown on the left- and right-hand sides of the figure. The two depict the sequence of activities in serving the caller and the destination, respectively. The *pots_server* functionality is shown in the center. The basic operation of the *pots_server* is as follows:

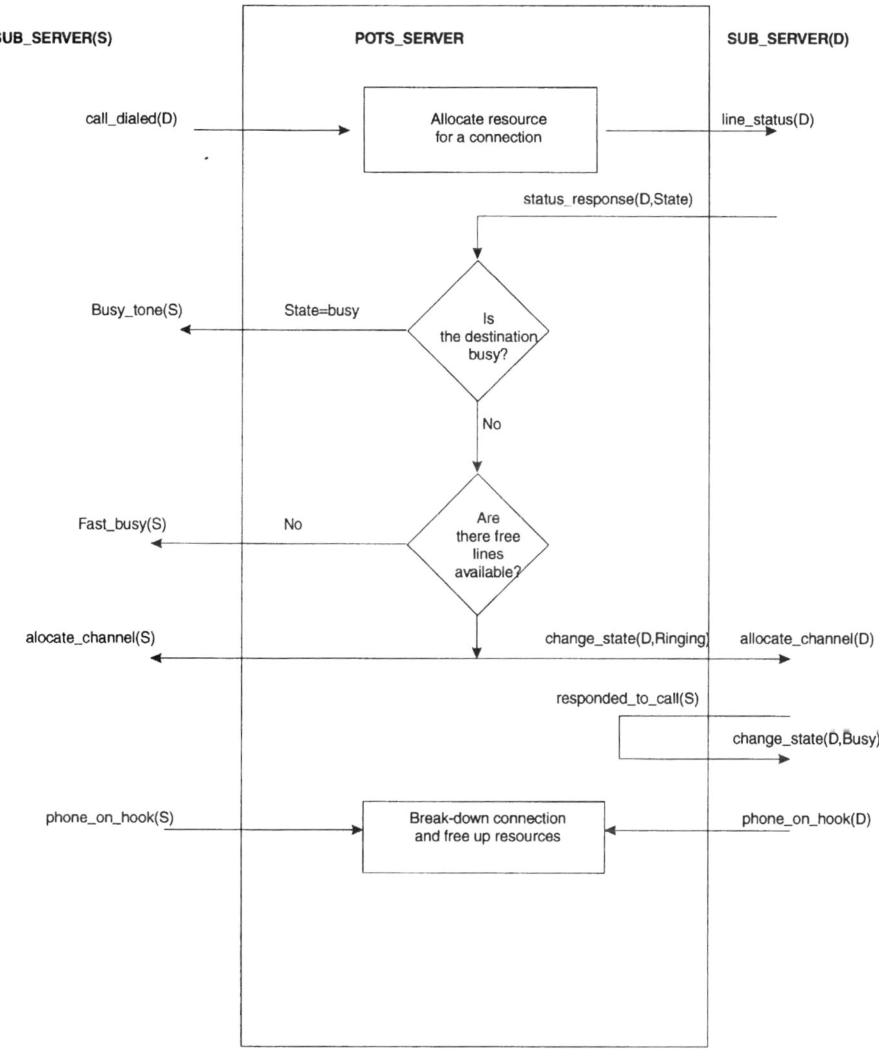

Figure 4.15 Service module interactions.

When a phone call is placed, a *call_dialed* message is communicated from the respective *interface_unit* of the *sub_server* module to the *pots_server* (within the *switch_element*), indicating the caller and destination phone numbers (the destination phone number in this example is identified by a "D," and the sender's by an "S"); resources are allocated by the *pots_server* to support the call setup. The *pots_server* then requests the status of the destination telephone from its *sub_server* with the *line_status(D)* message. The state is communicated to the *pots_server* through the *status_response(D,State)* message. A set of decisions is then made. If the destination is busy, a *busy_tone(S)* message is communicated back to the sender's *sub_server.* Otherwise, a check is made to determine if there is enough capacity left within the switch to establish the communication. If there is not, a *fast_busy(S)* message is sent back.

The *pots_server* must allocate the voice channels for the connection. A message (*allocate_channel*) is communicated to both *sub_servers* indicating the respective channels to use to establish the connection. Another message (*change_state(D, Ringing)*) is communicated to the destination to send a ring tone to the destination telephone. If the destination responds, a connection is made, a *responded_to_call(S)* message is sent to the *pots_server,* and the destination state is changed to busy. When one of the parties goes *on_hook* (signaled by a *phone_on_hook* message), then the *pots_server* breaks down the connection and frees up resources for other calls.

The behavior as specified within DSL follows the same message interactions. In addition, the resources and data structures for supporting the calls are identified. The *sub_server* maintains the state of the telephones connected to its *interface_unit* (see Fig. 4.16 for the *sub_server* code).

The *pots_server* module creates three resources to accommodate the call. The first is the *connect* resource, which identifies the two parties and the status of the connection. The status can be *connect_request* (connection not yet established), ringing (the destination is ringing), and connected (connection is established). The other two resources are the agent and the terminator, which contain local data for the caller and destination phone, respectively. Once complete, a message *line_status* is sent to the destination phone's *interface_unit* to determine if the telephone is occupied. The DSL code is:

```
(call_dialed(telno(ID, SDNO, STEL), telno(RSNO, RDNO, RTEL)):-
    create(connect(telno(ID, SDNO, STEL), telno(RSNO, RDNO, RTEL)),
    connect_request),
    create(agent(STEL), SDNO),
    create(terminator(RTEL), RDNO),
    send(_,line_port(RDNO), update_buffer(RDNO,
    line_status(RTEL))))
```

```
module(sub_server(ID, INPUT),
[
(channel_update(0, COMMAND):-send(COMMAND)),
(channel_update(CH, unused):- write('ignored data'),nl),
(channel_update(CH, VOICE):- check_res(connect_table(TEL), CH),
                             send(_, twisted_pair(TEL), VOICE)
),
(phone_off_hook(A):-
                      check_res(line_state(A), CALLSTATE),
                      send(resolve_call_state(A, CALLSTATE))
),
(resolve_call_state(A, onhook):-
                      set_res(line_state(A), offhook),
                      send(_, twisted_pair(A), dialtone)),
(resolve_call_state(A, busy):-
                      set_res(line_state(A), offhook),
                      send(_, twisted_pair(A), dialtone)),
(resolve_call_state(B, ringing):-
                      send(_, proc_port(ID), update_buffer(0,responded_call(telno(X,ID,B)) ))
),
(change_state(TEL, STATE):-
                      set_res(line_state(TEL), STATE)
),
(dial_dest(A, B):-
                      send(_, proc_port(ID), update_buffer(0,call_dialed(telno(X,ID,A), B)))
),
(phone_inactive(A):-
                      set_res(line_state(A), onhook),
                      send(_, proc_port(ID), update_buffer(0,phone_on_hook(telno(X,ID,A) )))
),
(line_status(A):-
                      check_res(line_state(A), STATE),
                      send(_,proc_port(ID), update_buffer(0,status_response(telno(X,ID,A),STATE)))
),
(channel_allocate(TEL, CH):-
                      set_res(connect_table(TEL), CH)
),
(fast_busy(A):- send(_, twisted_pair(A), fastbusy)
),
(busy_tone(A):- send(_, twisted_pair(A), busytone)
),
(voice_message(A, VOICE):-
                      check_res(connect_table(A), CH),
                      send(_, proc_port(ID), update_buffer(CH, voice_message(VOICE)) )
)
]).
```

Figure 4.16 *sub_server* module code.

As a response to the last message, the respective *interface_unit* will eventually send a *status_response* message indicating the state of the destination phone. The phone can be *on-hook* (not busy) or *busy* (talking to or dialing another line). In the former case, the *pots_server* will allocate a free pair of channels for the two telephones (this information is contained within a resource named *connect_table*) and also send ring tones to the two. If no free lines are available, a fast-busy tone will be sent to the calling party and the data structures will be cleaned up. The DSL code for the *pots_server* behavior is outlined in Fig. 4.17.

An example of some of the intricate message and resource manipulations can be observed in the code. It is worthwhile to walk through a segment of the code and highlight some of these. When the destination phone is picked up, a *responded_call* message is received by the *pots_server.* The module will consequently ensure that the proper data structures are updated. This is described by the code segment below:

```
(responded_call(telno(X,DSB,B)):-
      check_res(connect(A, telno(X,DSB,B)), ringing),
      check_res(terminator(B), Bport),
      set_res(connect(A,telno(X,DSB,B)), connected),
      send(_,line_port(Bport), update_buffer(Bport,change_state(B,
      busy))))),
```

The phone conversation will terminate when the originating party hangs up the telephone. The *phone_on_hook* message is generated by the respective *interface_unit* when a telephone has been detected as becoming on-hook. The *pots_server* determines whether the telephone is the originator or the destination and accordingly cleans and updates the respective data structures. This is reflected in DSL as four possible choices. The first checks to see if a connection is already made and if the sender has hung up. If so, the resources are freed as below:

```
(phone_on_hook(telno(X, DS,A)):-
    check_res(connect(telno(X, DS,A),telno(Y, DSB,B)), connected),
    remove(connect(telno(X, DS,A), telno(Y, DSB,B))),
    remove(agent(A), Aport),
    remove(terminator(B), Bport),
    set_res(channel_table(Aport, ChA), unused),
    set_res(channel_table(Bport, ChB), unused),
    send(_,line_port(Bport),
    update_buffer(Bport,phone_off_hook(B)))),
```

The second case is if the destination hung up. In this case, nothing is done until the caller goes on-hook. This is described simply by the code segment

```
(phone_on_hook(B):-
    check_res(connect(A,B), connected)),
```

If neither of the two cases above is satisfied, this implies that a phone went on-hook before a connection was established. Hence a check is made to determine if some resources were allocated by the module. If so, they are cleaned up as below:

```
(phone_on_hook(telno(SW,DS,A)):-
    check_res(agent(A), Aport),
```

```
/* TEL, REC = switch no, ds no, local no. - ie. 1,4,4567 makes up a phone number. */
module(pots_server(ID),
[
(responded_call(telno(X,DSB,B)):-
                check_res(connect(A, telno(X,DSB,B)), ringing),
                check_res(terminator(B), Bport),
                set_res(connect(A,telno(X,DSB,B)), connected),
                send(_,line_port(Bport),  update_buffer(Bport,change_state(B, busy)))),
(call_dialed(telno(ID, SDNO, STEL), telno(RSNO, RDNO, RTEL)):-
                create(connect(telno(ID, SDNO, STEL), telno(RSNO, RDNO, RTEL)), connect_request),
                create(agent(STEL), SDNO),
                create(terminator(RTEL), RDNO),
                send(_,line_port(RDNO),  update_buffer(RDNO, line_status(RTEL)))
),
(status_response(B, STATE):-
                check_res(connect(A, B), connect_request),
                send(called_state(A, B, STATE))
),
(phone_on_hook(telno(X, DS,A)):-
                check_res(connect(telno(X, DS,A),telno(Y, DSB,B)), connected),
                remove(connect(telno(X, DS,A), telno(Y, DSB,B))),
                remove(agent(A), Aport),
                remove(terminator(B), Bport),
                set_res(channel_table(Aport, ChA), unused),
                set_res(channel_table(Bport, ChB), unused),
                send(_,line_port(Bport),  update_buffer(Bport,phone_off_hook(B)))),
(phone_on_hook(B):-
                check_res(connect(A,B), connected)),
(phone_on_hook(telno(SW,DS,A)):-
                check_res(agent(A), Aport),
                remove(agent(A), Aport),
                send(_,line_port(Aport), update_buffer(Aport, change_state(A, onhook))) ),
(phone_on_hook(telno(SW,Aport,A)):-
                send(_,line_port(Aport), update_buffer(Aport, change_state(A, onhook)))
),
(called_state(telno(SW,DS,A), telno(BSW,BDS,B), busy):-
                check_res(agent(A), Aport),
                remove(connect(SW, telno(DS,A), telno(BSW, BDS,B)), SS),
                remove(terminator(B), Bport),
                send(_,line_port(Aport), update_buffer(Aport, busy_tone(A)))),
(called_state(telno(XX,DS,A), telno(YY, BDS,B), onhook):-
                check_res(agent(A), Aport),
                check_res(terminator(B), Bport),
                check_res(channel_table(Aport, ChA), unused),
                set_res(channel_table(Aport, ChA), A),
                check_res(channel_table(Bport, ChB), unused),
                set_res(channel_table(Bport, ChB), B),
                Index1 is ChA*20+Aport,
                Index2 is ChB*20+Bport,
                set_res(connect(telno(XX,DS,A), telno(YY, BDS,B)), ringing),
                send(_,mem_port(ID), mem_write(Index1, Index2)),
                send(_,mem_port(ID), mem_write(Index2, Index1)),
                send(_,line_port(Aport), update_buffer(Aport, channel_allocate(A, ChA))),
                send(_,line_port(Bport),  update_buffer(Bport,channel_allocate(B, ChB))),
                send(_,line_port(Bport),  update_buffer(Bport,change_state(B, ringing))),
                send(_,line_port(Aport), update_buffer(Index1, ring_tone)),
                send(_,line_port(Bport), update_buffer(Index2, ring_tone)) ),
(called_state(telno(XX,DS,A), telno(YY, BDS,B), onhook):-
                write('No free lines available'),
                remove(connect(telno(XX,DS,A), BB), SS),
                remove(terminator(B), Bport),
                send(_,line_port(DS), update_buffer(DS, fast_busy(A)))
```

Figure 4.17 *pots_server* module code.

```
remove(agent(A), Aport),
send(_,line_port(Aport), update_buffer(Aport, change_state(A,
onhook))) ),
```

If no resources were allocated, then the state of the sender is updated to *on_hook* by

```
(phone_on_hook(telno(SW,Aport,A)):-
    send(_,line_port(Aport), update_buffer(Aport, change_state
    (A, onhook))) ),
```

This completes the behavioral description of the telephone services. At this point, design exploration and further model development support constructs for DSL need to be established—which is the next topic.

Chapter

5

Design Execution

Like any modeling language, DSL embodies a philosophy of its own with regard to the design of systems. The language is suited to the top-down design approach, wherein design detail is added and further refined by inherited modules. An initial design commences with a main model consisting of a higher-order module. Subcomponents are then systematically defined or utilized to achieve a suitable level of design detail. The overall design is then subjected to a design execution phase, in which the model is simulated and different design trade-offs are examined. This approach requires a reasonable model library to support design exploration and a structured approach to the design (which enforces a modeling discipline upon the designer) (Booch 1991). This chapter will explore the issues related to design execution. The discussion will assume the existence of a hypothetical modeling environment which will support the notions presented therein. Chapter 6 will deal with the problems of implementing the hypothetical environment.

A complete DSL model suitable for execution was described as a composition of a model of the environment and of the system under consideration. The model of the environment represents a conglomerate of the external traffic and test cases to be applied to the model under consideration. Generally the applied load does not completely encapsulate all possible events that may interact with the system, but it is representative of known scenarios of interest. Test cases can be created through state-driven models, which can be described using DSL, or, alternatively, a model of the environment can generate them with another language (such as Prolog or *C*). Hence a mixture of DSL and other languages may be used. It should be noted that the model of the environment is used solely for experimentation and model support; it is never intended to be synthesized as part of the system.

Experimentation

DSL experimentation constructs consist of commands related to constraint manipulation and library activities. These commands are essential to the conduct of experiments and are described as

```
dsl_experiment_constructs:: = module_type_definition
          |constraint_definition
          |library_definition
```

The *module_type_definition* is specific to synthesis-oriented issues and will be treated in the next chapter; however, the latter two commands are addressed in this chapter.

Constraints

The *constraint* statement is a predicate which is useful during both design execution and synthesis. It imposes a restriction upon a module's resource or behavior. The form of the statement is

```
constraint_definition:: = constraint(module_name,entity,const_type,
values).
module_name:: = dsl_name
entity:: = dsl_name
const_type:: = dsl_name
```

where *module_name* is the name of the module which the constraint is applied to, *entity* defines what is being constrained (resource name or behavior name), *type* is the name of the type of constraint, and *values* is a constraint value applied to the *type.*

The label *type* is a user-defined computation of a constraint retained within the modeling environment. These can be hard-coded within Prolog or any other language. To illustrate the concept, assume that there is a desire to define a constraint *upper_limit* which will impose an upper bound for *module_name.* For example, *constraint(clock, Clock_rate, upper_limit, 100)* may indicate that the maximum clock rate for the indicated module is 100 (ns). The *type upper_limit* would correspond to Prolog code which computes the current value of *Clock_rate* and ensures that it is below the upper limit. The implementation details for the DSL statement are deferred to Chap. 6.

The *constraint* command is also used to identify any design violations during execution. Constraints are related to a module and are used to define limits upon parameters of a given module, resources, or messages. These can be defined to be systemwide constraints (affecting more than one module) or local constraints (restricted to individual modules). For example, the maximum size of a memory module can be regarded as a local constraint. However, if multiple instances of the memory module are used by a main DSL model to define a larger memory subsystem, then this can be viewed as a system-level constraint. System-level constraints are defined within the main DSL model, whereas local ones can be established within the library modules described below. After invocation of the library, the environment performs routine checks of the constraints so that they are not violated during simulation.

Analysis

DSL-based models permit the use of analysis techniques to prove different properties of the system. Analysis of DSL models can be described as being applicable in one of two cases:

- Properties that require analysis at a component module level
- Those that need analysis at a system level

Analysis capabilities differ significantly between the two cases. The first case is amenable to component-level analysis, where properties related to isolated modules are the focus. It would be desirable to perform such an analysis when a new module has been created or just before committing a module to a library. The second class of analysis is applied to all the modules simultaneously. Hence the point of interest at this level of analysis is in the interactions between modules or the side effects thereof. This would be best utilized during or after the actual simulation of the entire model. Use of system-level analysis before execution would most likely be premature, since the model (as viewed from a system level) may not be stable and may undergo further revisions.

From a simulation point of view, the two cases may be described as being equivalent to stages in the *verification* and *validation* of the model. These two terms are defined, from a simulation perspective, as follows:

- *Verification* is the act of determining whether or not subcomponents behave as expected or fulfill their anticipated requirements.
- *Validation* is an evaluation of the entire model (typically at the end of the simulation cycle) to ensure that its behavior is consistent with the requirements.

Hence the first case can be viewed as a verification of the constructed modules, while the second is the validation. These activities are viewed as part of the overall design execution process as represented in Fig. 5.1. The role of verification and validation within design execution is shown in the figure.

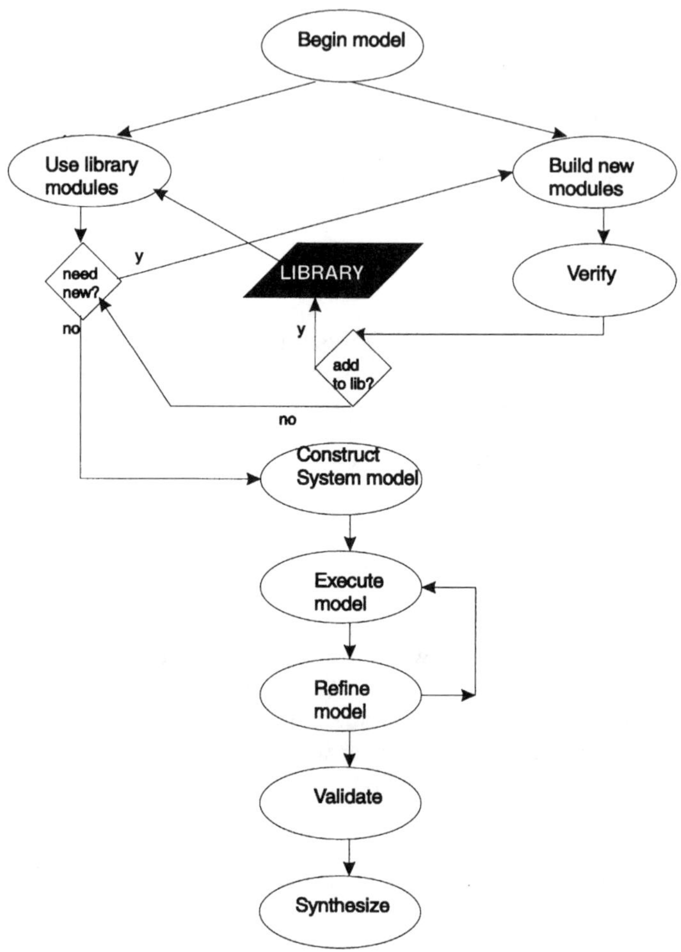

Figure 5.1 Verification and validation.

This section will illustrate possible analysis options available to the designer, using the predicate net formalism introduced in the earlier chapters. The analysis of large models requires a considerable effort by experienced modelers with a good understanding of the analysis methodology and the application domain. The techniques presented here are applications of basic analysis methods to the language's behavior as an introduction to DSL analysis. The innovative use of these techniques, along with the art of modeling and creativity (such as partitioning higher-order modules into subnets), can produce further analysis methods and benefits.

Module verification

Module verification restricts the analysis to individual modules. The corresponding predicate net is subsequently limited to the behavior representation of the modules. This limits the overall complexity of the number of reachable states of the underlying predicate net. Representations of isolated modules are strict nets (one message is processed at a time), and consequently the number of tokens in places is manageable. A module's representation can be analyzed using reachability or invariant techniques that are commonly applied to predicate nets (see Genrich 1981 for a formal treatment of net analysis). Module analysis is applied to determine:

- *Inconsistent behavior definitions.* The potential for writing incorrect behavior code is always present. These types of descriptions can lead to infinite loops, where behavior execution is never completed. Alternatively, a behavior may introduce an unwanted state into the module.
- *Unused DSL code.* Code which serves no purpose or never invokes any state changes can be detected.
- *Undefined message ports.* Ports which do not exist can be deduced from the syntax of the *send* statements in the module's behavior.

Some basic properties of modules can be verified at this level of analysis. These include safeness, sequence of firings, reachability to a particular state, and liveness.

Safeness. The semantics of its behavioral execution implies that a module's net representation is safe, i.e., the Petri-net representation of a module will have no more than one token in any given place. Subsequently, any corresponding net representing a module's operation must result in a safe net. To perform the traditional analysis of a net generated from module behavior, the desired initial markings must be specified. The net will have some tokens placed as part of the starting conditions of a module. However, there is a need to identify the incoming token stream (representing the possible messages). This involves playing the token game. The Petri-net can be initialized in many different ways, depending upon the analysis objective. One possible scenario is illustrated in Fig. 5.2.

The PrTN representation of a module is indicated by the shaded region. Around this representation, a supporting net can be attached consisting of a place to hold all the messages that can be interpreted by the module, a place (busy) that ensures that only one message is processed at a time, and a place (message) that identifies the message being processed by the PrTN. The generation of the supporting net is

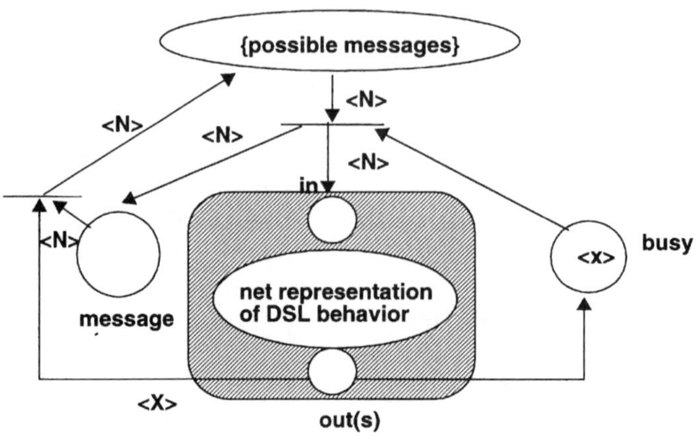

Figure 5.2 Module analysis configuration.

trivial, since only three places and two transitions are required. The input place to hold the incoming messages can be populated with tokens typed as possible message names (extracted from the module behavior). Consequently, a reachability tree can be generated with a variety of existing Petri tools, and analysis of properties pursued.

Reachability. An example of a timer was introduced in Fig. 4.10. Expanding upon this example, the reachability tree for the predicate net of the timer is shown in Fig. 5.3. The initial marking assumes either a *reset* or *clk* type token arriving at the *input* node (no other message type is acceptable). In the event that a *reset* token arrives, the resource value (within *res_count*) is initialized to 0. The *clk* token causes the value to increment or a reset to occur. The latter takes place when the resource is equal to the count limit defined by *Value*.

This graph is small and can possibly be scrutinized by manual means, but in general, analysis tools are necessary. A multitude of tools exist for performing reachability analysis of Petri-nets.

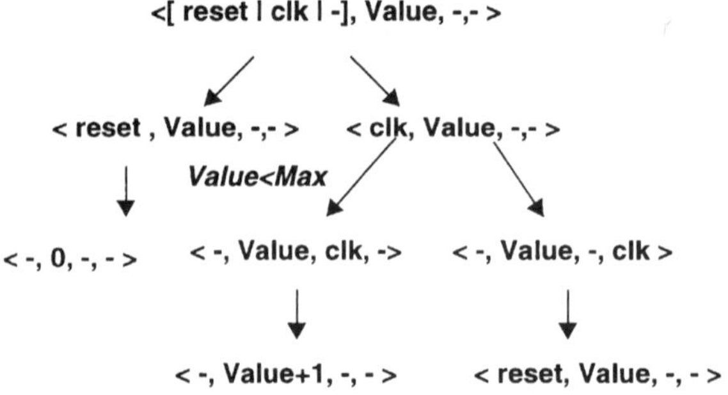

Figure 5.3 Timer reachability tree.

Liveness. Liveness analysis pertains to detection of possible markings which can never change. An indication of liveness is the presence of a marking at a leaf node in a reachability graph which has no subsequent defined marking. In the case of modules, this may reflect an ill-defined segment of the behavior where nothing useful is accomplished by a message. A consequence can be a module which can become indefinitely suspended, or one where a given behavior performs no new operations.

System validation

System validation consists of techniques applied to aggregate models or higher-order modules. The treatment of system-level models is not as straightforward as that for component modules. System models, although composed of component modules, do not generally exhibit a safeness property. This becomes apparent when it is observed that there are implicit queues within each module. When these properties are translated to nets, bounds on places may not be known beforehand. As a result, models can contain unbounded places and exponentially increasing state spaces, making analysis very difficult.

The objectives of analysis applied during the validation phases are significantly different from those for verification. Some of the properties of interest may be:

- The validation of performance bounds of the model. Average performance metrics are extracted from analysis and compared to the design requirements and specifications.
- The validation of buffer requirements for different components. Analysis of queueing times at different parts of the model can be used to validate the system.

Different approaches to examining the system model can be taken, depending upon the desired emphasis. The approach that will be presented is based upon a simulation of a network of queues. In reality, it is an approximation of more detailed interactions that exist between modules. However, it provides sufficient granularity of detail to obtain results for analysis. Depending upon the complexity of the model, a more traditional queuing-based analysis can also be performed by carefully applying assumptions to the model (Kleinrock 1976, Ross 1992).

A simulation-based queuing approach views each module as a single-server first-in–first-out (FIFO) queue with multiple service times. The number of service times can be derived from the number of different message headers contained within the module's behavior. If a module is capable of responding to five different message types, this implies five service times. The service times of the queues are derived from the delay parameters associated with the message. There are also routing probabilities attached to the queue. A significant amount of the overall accuracy of the queuing model is attributed to all of the probabilistic parameters. They can be defined using experience and educated guesses or with the help of the underlying simulation statistics. Queuing statistics can be gathered while the model is executed to help define first assumptions of the probability parameters. The difficulty of constructing an approximate model resides in establishing meaningful service times for the queues. This is best illustrated through a progression of some simple examples (these are depicted pictorially in Fig. 5.4).

Consider a simple behavior described in a module as:

```
m1:-delay(d1), send(_, _, ma).
m2:-delay(d2),send(_, _, mb).
```

The module accepts two message types (*m1* and *m2*) and will, after a delay of *d1* or *d2* time units, send message *ma* or *mb*. To represent this situation in a queue-based model, we assume that an inbound probability P_{mi} is given. This represents the probability of a message *mi* arriving. The queue's service time is dependent upon which message has arrived. An approximation of the service time *D* can be given as

$$D = (P_{m1} . d1) + (P_{m2} . d2)$$

which is a weighted random choice between *d1* and *d2*. After the respective service time, a message is sent to one of two destinations (given by *A* and *B*). This can easily be represented in a queuing-based simulator.

Now let us consider a more complex situation in which messages are subject to conditional statements:

```
m1(true):-delay(d1), send(_, _, ma), send(_, _, mb).
m1(false):-delay(d3).
m2:-delay(d2), send(_, _, ma).
```

In this example, the module can accept a message *m1* and, subject to a condition (*true* or *false*) being satisfied, can potentially send two concurrent messages *ma* and *mb*. It can also interpret a message *m2* and consequently send a message *ma*. The case for message *m1* illustrates the problem with dealing with conditional constructs. Conditionals can be triggered in a multitude of ways. The example shows a testing of a true-false relation, but it could also have been a check for a value of a resource. This makes identification of the probability of executing the first or second behavior for *m1* difficult. However, for

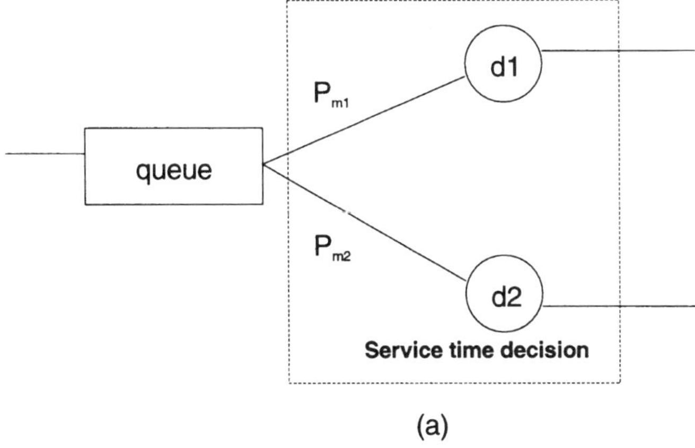

(a)

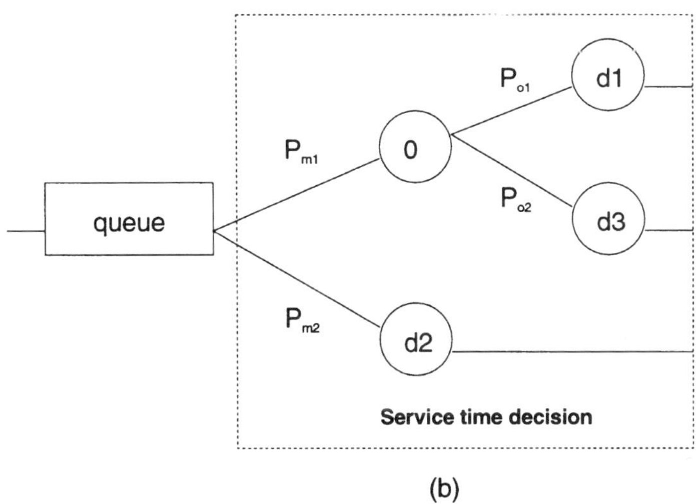

(b)

Figure 5.4 Examples of queue representations.

the time being, let us assume that the probabilities of executing the first and second behaviors for *m1* are known and given by *Po1* and *Po2*, respectively. Then the probability of the service time being *d1* is given by $P_{m1} \bullet Po1$, and that of its being *d2* is given by P_{m2}. Finally, the probability of *d3* is given by $P_{m1} \bullet Po2$. Again, these parameters can be programmed into a simulator.

The probabilities for the above queues can be obtained as follows:

- To obtain the inbound probabilities, the number of arriving messages for each module are cumulated and defined as m_{in}. A count for each message name m_i is also maintained as $m_i(k)$. These data can be gathered through the DSL simulator (defined during simulator configuration). After the end of the simulation scenarios, each probability is defined as the inverse ratio: $P_{mi} = m_i(k)/m_{in}$.
- Assigning probabilities due to conditionals (such as P_{o1} and P_{o2} above) may be more difficult. Some insight may be gained by monitoring the messages sent from each module and their respective counts. Some correlation may be possible to identify probabilities for conditional constructs. For example, if messages generated by the execution of the messages within a module are all unique (this implies that all message name and destination pairs used by the module are unique), then probabilities can be established by relating them to the total number of messages sent. Alternatively, a DSL probe can be placed within each conditional statement to gather statistics on the probabilities. However, this is costly in terms of the simulation time for the DSL model.

A network of queues can be constructed which represents an approximate model of the overall system behavior. Given that the modules are represented as queues, statistics obtained through queue analysis would help understand the system-level model metrics.

Library

The power of DSL and any other architectural-design language is in its ability to support design execution under the auspices of a solid library system. An initial DSL model can undergo several refinements during the exploration of different design alternatives. Figure 5.5 depicts the interactions between the design environment components and the DSL model. The initial model is at a very close level of abstraction to the design requirements; hence, if the requirements are not specific, the DSL model will leave room for significant refinement. In such circumstances, the user may intervene to make design decisions to help narrow the design space. Refinement occurs during the course of simulation as a consequence of interactions between the environment and the user.

DSL has a set of library commands which support library functionality. These commands are executed once during invocation. The library support commands for DSL are defined as:

```
library_definition:: = use_dsl_library([dsl_name*]).
            |configure_library_rules
            |start_cond_rules
```

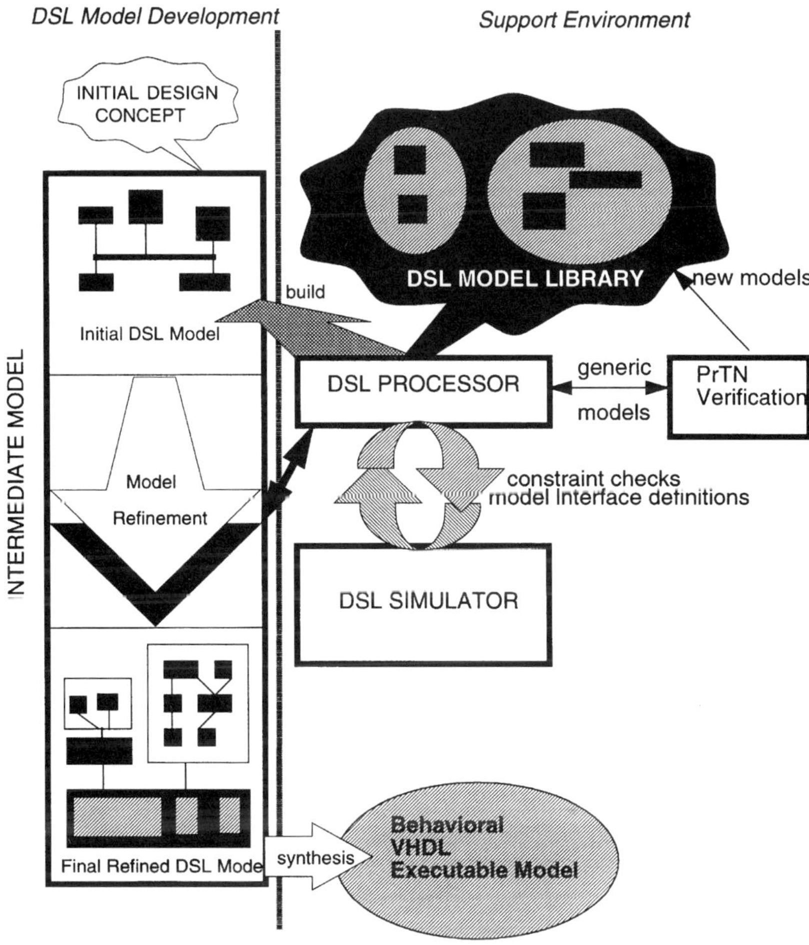

Figure 5.5 DSL and the modeling environment.

```
configure_library_rule:: = configure_library(dsl_name):- {Prolog
rules and DSL statements},
start_cond_rules:: = start_cond(dsl_name):- {Prolog rules and DSL
statements}.
```

The *use_dsl_library* command references library module names (*dsl_name*) to be used within a given DSL program. A DSL program is free-form, but adheres to basic Prolog syntactical guidelines. The two additional commands are needed to automate the setup of DSL models. They are both headers for Prolog rules (and consequently DSL statements) to be used to describe the actions to be taken upon the respective modules. These concepts are further elaborated below.

Configuration rules

The *configure_library_rules* are rules associated with specific (library) modules. They are intended to describe the initial values that the module may need upon initiation. The possible types of actions that can be undertaken by the rules fall into three categories:

- Rules to instantiate submodules
- Rules to instantiate resources
- Rules that define default interconnections

These categories of rules utilize powerful elements to facilitate the design and retrieval of generic reusable library modules. All the rules are assertions. This implies that either DSL elements are asserted (created within the Prolog environment) or other Prolog predicates are utilized in the evaluation of the rule. In either case, the *configure_library* statement corresponds to a Prolog rule which always evaluates to a *true* assertion. None of the elements in the body of the rule may be false.

A powerful tool when creating configuration rules is *parameterization.* This concept embodies the simple idea that parameter passing between modules can facilitate the definition of the structural description of the entire model upon invocation of the library modules. If variables are assigned to module parameters and are bound to specific types and values only during invocation, then they can structurally adapt their interface to the parent model. An example of the power of parameterization is illustrated in Fig. 5.6.

The figure depicts the use of a single parameter to generate and interconnect identical modules. The interconnection scheme is that of a 16×16 Banyan network used as a switching fabric for some asynchronous transfer mode (ATM) switches. The parameter used in constructing the network is the fabric size (in this case, 16). The interesting point in this example is that messages arriving at any input line of the fabric can be routed to any other output line. The figure depicts a four-stage configuration in which each stage consists of eight modules named *nodes.* Messages arriving at the inputs of the fabric are composed of a header (a binary-encoded address uniquely identifying one of the output lines) and the message body (irrelevant for the purposes of this example).

Each module can route a message to one of two lines based upon the value of the most significant bit of the destination header. If the bit is a 0, the message is sent through the upper output line of the module. If it is a 1, it is routed to the other output. At each stage, the destination header is left-shifted so that the most significant bit from the previous stage is discarded. It should be noted that for a fabric of

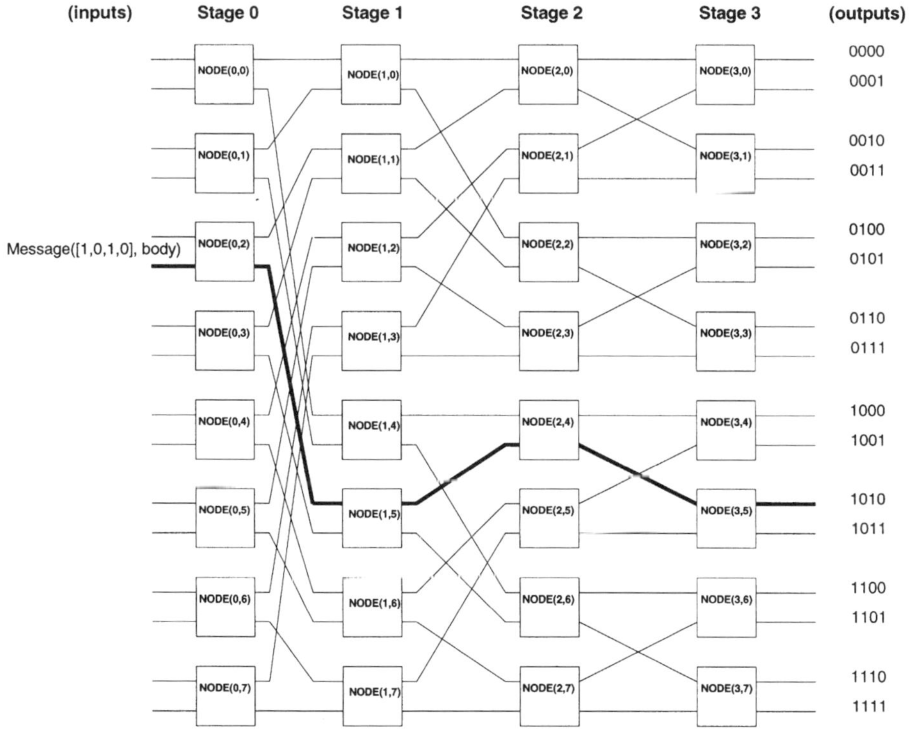

Figure 5.6 Self-routing switch fabric.

size M, there will be $\log_2 M$ stages, and the header would also require an equivalent number of bits to route the messages. Each stage will also contain up to $M/2$ modules.

The figure also illustrates the movement of a sample message destined for an output port (1010). The highlighted connections indicate the path that would be taken by the message. It is left as an exercise to verify that the destination will be reached by the message irrespective of the starting input port.

An algorithm to generate the interconnections for the modules is given below:

```
Stage = log2(M);
k = M/4;
For S = 0 to (Stage-1) do
Begin{
    For L = 0 to (2^S-1) do
    Begin {
        For ii = 0 to (M/(2×2^S)-1) do
        Begin {
            i = ii+L×k;
            j = floor(ii/2)+2×L;
```

```
            create a path: path( node(S, i), node(S+1, j), [ out(0),
            in(y) ]);
            create a path: path( node(S, i), node(S+1, k+j),
            [ out(1), in(y) ]);
        end (ii loop)}
    end (L loop)}
    k = k/2;
end (main loop)}
```

This algorithm will generate the interconnections for any size switching fabric. The modules assume that the output ports are named *out(0)* and *out(1),* whereas the inputs are *in(y).* The labeling for the inputs is not necessary and can be omitted unless the module behavior specifically requires the input port labels. The algorithm requires one parameter (the switch size *M*) to create all the module interconnections. After instantiation of the nodes, the algorithm is used to establish the interconnections.

Parameterization enforces a high degree of reuse in the design. Parameterized elements of the library modules are bound to variables as late as possible (when they are required by the parent module rather than during declaration), enabling their use in a multitude of different scenarios. Any variables are bound during the design execution. If some variables are not appropriately defined at this point, module analysis (see below) will detect the inconsistencies. The environment or user may then take appropriate action. The details of how modules are instantiated and interconnected through configuration rules are presented next.

Submodule instantiation. Submodule instantiation is a common use of configuration rules. A library module which refers to other library modules will use the configuration rules to instantiate the modules it requires. Library modules are invoked with the *use_dsl_library ([NAMES])* DSL statement. This makes available the behavior of the referenced library modules. The rules can then generate instances of the referenced module. This is achieved by incorporating the DSL *isa* statement into the rule. Since all the statements are essentially Prolog entities (or any other language if it is the implementation language), then recursion can also be used to generate multiple instances of modules.

An example of submodule instantiation is provided in the code of Fig. 5.7. The code depicts a higher-order module defined as *main* consisting of submodules *subscriber* and *residential_switch.* The example depicts a possible experimental setup for the DTSS example which was introduced in the previous chapters. *Main* also has four parameters associated with it depicting the number of telephones to connect to *residential_switch,* an identification for the switch, number of channels per interface card, and the number of interface cards.

```
ho_module(main(10, 0, 5, 2), [subscriber(X), residential_switch(X)]).
/* Main parameters are: no of telephones, switch no, no of channels/card, no of int. cards */

use_dsl_library([
                          dts_switch,
                          telephone
                          ]).

isa(subscriber(X), user(X)).
path(subscriber(T), residential_switch(SW), [tel_line(T), tel_line(T)]).

configure_library(main):-
          ho_module(main(T, SW, CH, CARD), XXX),
          asserta(isa(residentiall_switch(SW), dts_switch(SW, T, CARD, CH))),
          asserta(ho_module(residential_switch(SW), [])),
          configure_tel_con(T, SW).


configure_tel_con(1, SW):-
          asserta(module(subscriber(0), [])),
          write('Telephones connected to the switch...'),nl.

configure_tel_con(T, SW):-T>1,
          TT is T-1,
          asserta(module(subscriber(TT), [])),
```

Figure 5.7 Submodule instantiation example.

The first statement in the module description is the *use_dsl_library*. This determines that three other library modules are referenced by *main: dts_switch, telephone,* and *user*. The behavior of the three is irrelevant at the moment, but how they must be configured by *main* is important. In the example, a simple instantiation statement defines the relationship between *subscriber* and *user* as *isa(subscriber(X), user(X))*. Hence *subscriber* is an instance of the library module *user*. The line after the statement interconnects the *subscriber* to the *residential_switch*.

Even though the instantiation of *subscriber* was relatively straightforward, it is not a complete definition in terms of DSL specification. The description is valid for a generic module, but in this case, *main* has determined the number of subscribers that will be modeled (1000 in the figure). Hence each *subscriber* definition must be available; otherwise the environment will have no knowledge of what type of DSL entity *subscriber* is (module versus higher-order module). This is where the configuration rules are used.

The configuration rule first defines the higher-order module *residential_switch,* and then defines an inheritance relationship from the *dts_switch* library module. To generate the *subscriber* module definitions, a recursive Prolog predicate *configure_tel_con* is used. The predicate generates 1000 subscribers starting from *subscriber(0)*. Note that the number of subscribers is affected only by the parameters defined by the *main* higher-order module. Hence, another simulation with more or fewer subscribers would imply simply a parameter change in *main*. All other configurations for *main* would be transparent to the modeler.

Resource instantiation. Following an approach similar to that for submodule instantiation, resources can also be defined within configuration rules. This becomes convenient when the number of resources or the default values are dependent upon how the parent module is configured within a model. For example, if the number of channel buffers in a switch depends upon the number of subscribers connected to the switch (remember that buffers can be hardware or software entities), a rule is needed to determine how many buffers are to be created when the switch module is invoked.

Resources are defined within library modules and instantiated through their invocation by a parent module. There is a set of activities that must be followed to define resources. For example, a library module M_1 referenced by a higher-order module M_{parent} (M_1 would be contained in a *use_dsl_library* statement within M_{parent}) would have an associated set of resources R_1. These resources may have been explicitly defined in the library module or required by its behavior. The parent module contains the instantiation statement ***isa(M**$_i$**, M**$_l$**)***. This describes the relationship between the instance (M_i) and the library module. Since this instance is already defined before the library module is invoked, the library module's configuration rules can utilize this to generate the resources for the various instances.

The configuration rule would resemble

```
Configuration_rule(m_parent):-
        isa(MODULE_INSTANCE, m_l),
        asserta(resource(MODULE_INSTANCE, name_of_resource,
        value)).
```

The parameters *m_parent* and *m_l* refer to the parent and library module names, respectively. The resource is defined to consist of a label called *name_of_resource* with an initial value *value*. In this case, only one resource is required for each instance of the module. Additional resources could have been defined using additional Prolog *asserta* predicates or using another predicate to define a recursive operation (as with submodule instantiation) to generate a large set of resources. If some operations are prevalent within a given design group, recursive rules governing the generation of the resources could be saved and reused as part of the standard design practices.

Module interconnection. Module interconnection is defined through the use of *path* statements. The inclusion of *path* statements within configuration rules identifies predefined communication paths between the library module and its component modules. This does not imply that other paths cannot be allocated by the user or environment during design execution. The implication is simply that the interconnections establish default paths that are deemed useful in facilitating communication between the modules.

As with resource instantiation, *path* statements are created using the Prolog *asserta* predicate. Similarly, recursive rules can be utilized to generate various interconnection schemes with little work. Three common interconnection descriptions are given in Fig. 5.8. The first depicts a single path connection between two types of modules. The modeling environment would have access to only one path to communicate between the modules, after which it would need to generate additional paths of its own (during run time). To establish this type of configuration, the path statements would resemble

```
path(module_A(X), module_B(X), [a, b]).
```

The second scenario is that of one-to-many communication. In this case, there are several alternative paths for establishing communication. In this case, the *path* statement is

```
path(module_A(X), module_B(Y), [a, b(X)]).
```

The last example depicts a one-path-to-many-module interconnection. If a communication does not successfully occur with one path, the alternative is to try another module. The *path* statement for this configuration is

```
path(module_A(X), module_B(Y), [a(X), _]).
```

The example of the Banyan ATM switching fabric introduced in the preceding sections can now be described in terms of its DSL representation. The configuration rules for the switching fabric module are given in Fig. 5.9. The configuration rule, after instantiating the node modules (omitted from the figure), initiates a set of recursive Prolog predicates through the *interconnect_nodes(Stage)* predicate. The predicate implements the main *for loop* of the algorithm by recursively executing a set of predicates which represent the inner loops (*loop1* and *loop2*). The innermost loop (*loop2*) creates the DSL interconnections between modules by adding the respective *path* statements to the Prolog database. As expected, the parameter required to establish

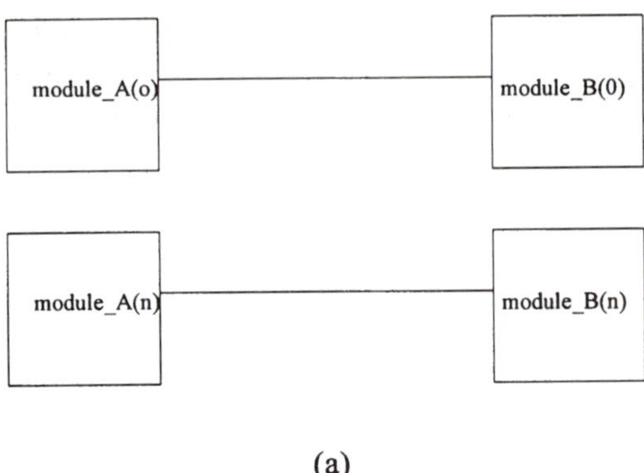

(a)

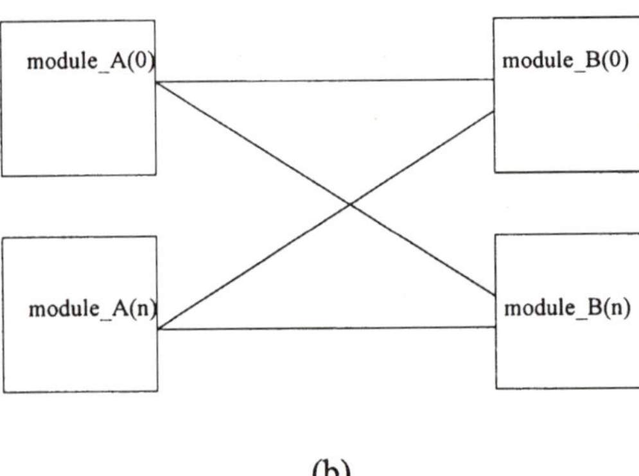

(b)

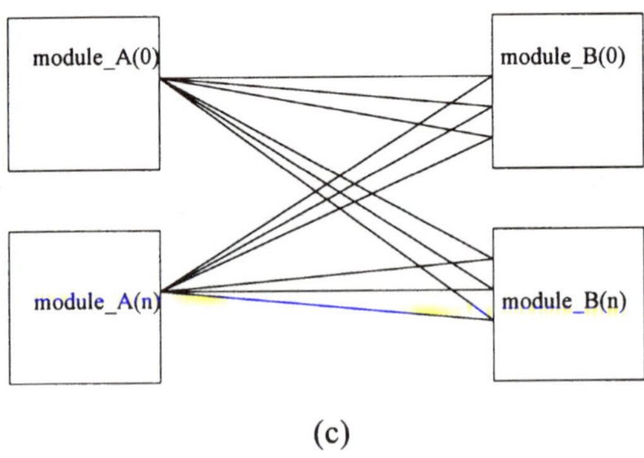

(c)

Figure 5.8 Module interconnections.

```
ho_module(switching_fabric(SIZE), [ node(S,I)]).
...
Congiration_rule(switching_fabric):-
{rules to instantiate the nodes},
isa(_, switching_fabric(SIZE)),
Stage = log(SIZE),
interconnect_nodes(Stage).

interconnect_nodes(-1).

interconnect_nodes(S):-
Loop1_limit = (2exp(S)-1),
loop1(L, S, K),
interconnect_nodes(S-1).

loop1(-1, S, K).
loop1(L, S, K):-
I = (M/(2x2exp(S))-1),
loop2(L, I, S, K),
loop1(L-1, S, K).


loop2(L, -1, S, K).
Loop2(L, I, S, K):-
i = I+2xLxK,
j = floor(I/2) + 2xL,
asserta(path(node(S, i), node(S+1, j), [out(0), in])),
asserta(path(node(S, i), node(S+1, K+j), [out(1), in])),
Loop2(L, I-1, S, K).
```

Figure 5.9 Configuration rule for switching fabric.

the connections is derived solely from the fabric size (defined as *SIZE* in the figure). Consequently, modules requiring the library module *switching_fabric* need only communicate the switch size.

Start conditions

The *start_cond_rules* refer to the initial conditions required by the module to effectively represent its behavior. The objective of the rules is to place the modules in a known state. A number of possible rules exist, as described below.

Message initiation. When modules are first brought into a model, they are essentially passive entities in the sense that they wait to respond to outside stimuli. This may be suitable for some scenarios, but there are cases in which the module needs to be active right away. For example, the clock that was given in Fig. 4.12 is a case where the module needs to be immediately involved with other modules. The clock must be placed into a state where it can start its count sequence irrespective of other modules. In fact, the clock module generates messages without any external stimuli; hence, if not activated, it would never generate anything and would be of no use to other modules.

The start conditions permit message initiation, which in turn permits modules to be activated by any sequence of messages. In the case of the clock behavior, this can take the form of

```
start_cond(generic_clock):-isa(X, generic_clock(_, NO)),
send(clock_count(NO)).
```

The header of the rule implies that the rule is associated with the *generic_clock* library module. The DSL *isa* statement implies that the rule is applied to all instances of the *generic_clock*. In addition, the parameter *NO* is used to define the upper count limit for the clock (refer to the DSL description in Fig. 4.12). The final statement generates a *clock_count(NO)* message to all instances of the module. This message will be executed immediately by the module(s) when the simulation commences. In this case, the clock behavior will generate a continuous periodic stream of clock messages which can be interpreted by the destination modules.

There are a few subtle issues with message initiation that are resolved at the environment level. The first issue is the application of the start conditions to the instances of the library module. The library module provides a template of the behavior which is inherited by the instances. Hence it should be observed that the instances, not the library module, require the initial messages. Consequently, *send* statements within the *start_cond* rules will almost always be proceeded by an *isa* statement. In cases where this is not so, the environment must guarantee that the messages are applied to the corresponding instances.

The second issue is that the environment needs to establish a mechanism for placing the initiated messages on the simulation queues so that they are processed. This is not a major difficulty, but will vary according to how the simulator is implemented.

Constraint definition. In addition to constraints defined by the user during the setup of an experiment, constraints can also be part of a library module. A constraint is treated like a resource and initiated in the same manner. Through parameterization, nontrivial constraint-oriented dependencies between modules can be established.

Observation hooks. DSL has no predefined constructs related to observation of data or events within a model. Naturally these are essential features for any simulation language; however, they are readily implemented using the underlying language. By default, DSL does not collect any type of data. This permits users to selectively determine the amount of data to collect from simulation runs. Since data collection can consume a significant amount of processor time, which can translate to long simulation times for large models, minimizing the amount of data collection can be highly desirable.

There are two types of data that are observable in DSL simulations. The first kind is a collection of statistics of system activity. This includes message delays, queuing times, and module utilization. These are simulator-controlled data sets and can be included as part of the simulator semantics.

The second type of data pertains to those used in verification of the functionality and behavior of the system. The activities related to the data consist of, but are not limited to, the detection of message arrivals, activation of an event, and change in resource values. These are all indicators that can help verify some aspect of the system behavior and are analogous to debuggers used for coding. In this sense, they can provide a trace of different simulation events for the purpose of validating their functionality. The method for observing the data can be included as part of the module description (such as *write* statements in the module behavior) or as part of the simulator. Because of the potentially large number of events and state changes that can occur, as well as the level of detail that may be covered by the model, selectively defining observation points is helpful.

A possible way of managing the complexity of observing events is to exploit the implicit hierarchy in the library structure. Figure 5.10 depicts an example of the library hierarchy for a hypothetical model of a system. The root of the tree is the most abstract level of the model. In the figure this is a higher-order module named *experiment*

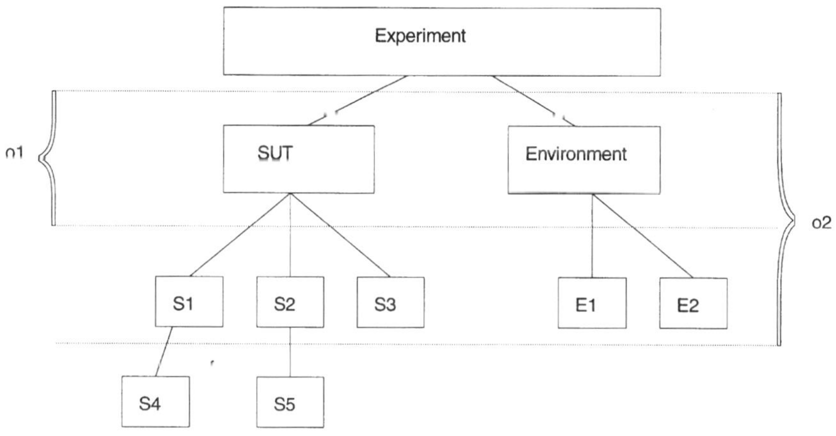

Figure 5.10 Model hierarchy for DSL.

and is composed of a module representing the system under test (*SUT*) and the *environment.* All the leaves emanating from the *SUT* are library modules which are required by it to depict the necessary structure and behavior of the entire system. At the leaves of the resulting tree are modules that do not inherit any behavior but their own.

Movement from one level of the tree to another can be viewed as observing more detail. In other words, the root of the tree is the most abstract description of the system, while the leaves are the most detailed. In relation to observing events, this stratification based upon the library inheritance hierarchy presents a useful mechanism for collecting data on pertinent events. If the simulator can associate the occurrence of events (such as a message arrival) at a module with a particular level in the tree, then restricting data collection to events at a given level can be possible. This is shown as o_1 in the figure. Furthermore, a range of levels can also be identified permitting observation of events in the range (shown as o_2).

These *observation levels* can be created by the simulator during initial configuration.

DTSS example revisited

The DSL model for the DTSS that was introduced in the preceding chapters can be structured as a reusable library model. Library structure is maintained upon invocation as a tree of modules, where the root is the topmost level (level 0) module which has directly or indirectly instantiated the others. Each parent library module configures its sibling within the environment as required. This property is demonstrated in Fig. 5.11, which shows a segment of DSL and user-written Prolog code which is part of the *dts_switch* library module's configuration rules.

The *dts_switch* module (the parent module in this case) identifies the sibling library modules that it requires using the *use_dsl_library* statement. The next statement, *configure_library(dts_switch),* defines the configuration rules for resources, ports, and module instantiation that are identified by the *dts_switch* module. In this case, the *generic_switch_element* and *interface_unit* modules are instantiated using configuration parameters (such as number of input/output lines and channels) defined from the instance of the *dts_switch* module. A subrule, *example_config,* determines the number of *interface_units* to instantiate and also defines the port interconnections for each with the *dts_switch.*

The effects of the library configuration rules are demonstrated in Fig. 5.12. The figure shows an example of a *residential_switch(0)* mod-

```
ho_module (dts_switch(NUMBER, INPUTS, CARDS,
      CHANNELS), [switch_net(NUMBER), interface_card(Y)]).

use_dsl_library([
            generic_switch_element,
            interface_unit
            ]).

configure_library(dts_switch):-
            isa(INSTANCE, dts_switch(N,I,C,CH)),
            II is I//C,
            asserta(isa(switch_net(N), generic_switch_element(C,CH))),
            asserta(isa(interface_card(Y), interface_unit(II, CH))),
            assérta(path(interface_card(Y), switch_net(N), [into_port(Y), trunk_in(Y)])),
            asserta(path(switch_net(N), interface_card(Y), [trunk_out(Y), inti_port(Y)])),
            asserta(ho_module(switch_net(N), [])),
            example_config(N, II, C).

example_config(N, IN, 1):-
            Index is N*16,
            asserta(ho_module(interface_card(Index), [])),
            tel_port_config(Index, IN, 0).
example_config(N, IN, CARD):-
            CARD1 is CARD-1,
            Index is N*16+CARD1,
            asserta(ho_module(interface_card(Index), [])),
            tel_port_config(Index, IN, CARD1),
            example_config(N, IN, CARD1).

tel_port_config(Index, 0, CARD1).
tel_port_config(Index, IN, CARD1):-
            isa(INSTANCE, dts_switch(N,I,C,CH)),
            IPRIME is I//C,
            INN is IN-1,
            Index2 is IPRIME*CARD1 + INN,
            asserta(path(interface_card(Index), INSTANCE, [t_pair(Index2), tel_line(Index2)])),
            asserta(path(INSTANCE, interface_card(Index), [tel_line(Index2), t_pair(Index2)])),
            tel_port_config(Index, INN, CARD1).
```

Figure 5.11 Configuration code for *dts_switch*.

ule which is an instance of the library module *dts_switch(0, 20, 2, 4)*. The parameters identify the *residential_switch(0)* to be configured as a switch with a switch number 0, 20 telephone lines supported, 2 interface units desired, and 4 channels per multiplexed line. The parameters have been kept small so that their effects can be captured in the figure. There is also a significant impact upon the resources within the various modules. Some of the major modules that are affected are shaded.

The restriction of the design domain enables the use of library modules with different configuration rules in various interconnection schemes. This is useful to enhance model modularity and reusability within the application domain.

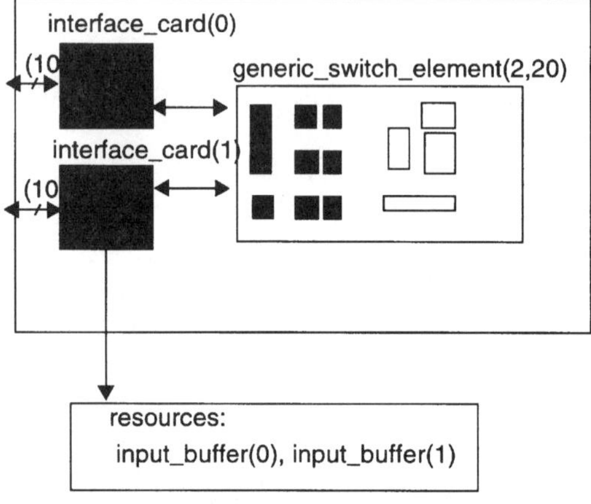

Figure 5.12 DSL diagram for result of configuration rules.

Anatomy of a DSL Simulator

The fundamental component of the run-time environment is the simulator engine. The engine to execute DSL models follows the design of discrete-event-oriented simulators. The architectural-level design execution issues that have been presented also add an extra dimension to the simulator design. The dissection of the simulator is shown in Fig. 5.13.

The figure is partitioned into two distinct components. The first is the DSL interpreter. This component interprets the DSL model and executes the relevant DSL statements. As a result, the underlying model may undergo several changes, causing the creation of new messages or modification of resources. The interpreter has two roles to play. The first role is to act as an interface to the library system. When the DSL model is first executed, the interpreter needs to access and configure the relevant library modules. The second role is the actual execution of the model. This implies that DSL statements are interpreted and their actions appropriately undertaken by the interpreter.

The other component in the figure is the scheduler subsystem. The scheduler uses an event list to determine which event is to be proces-

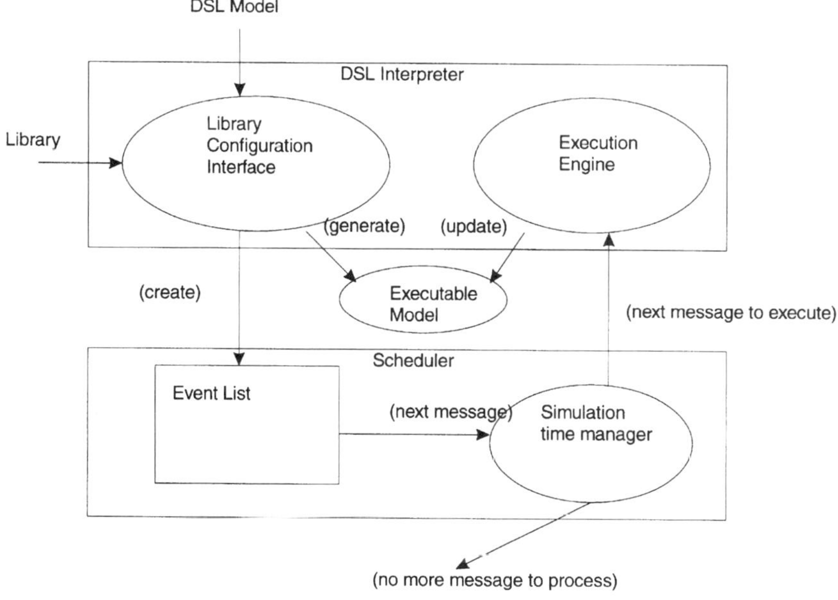

Figure 5.13 Anatomy of a simulator.

sed. This evaluation is made by referencing a simulator clock. The event is then removed from the list and sent off to the processor. Integral to the scheduler's performance is the management of the event list and simulator clock.

A variety of scheduling approaches are possible for discrete event simulators (Fishwick 1995). In reality, a scheduler is serializing a highly parallel and concurrent set of model interactions so that it is usually suitable to be executed on a single processor based upon the traditional Von Neumann architecture. A DSL simulator must facilitate the execution of messages, since these are the primary means for initiating a change in the state of the system. The arrival or generation of a message at or from a module can be interpreted as an event. Hence it appears natural to design an *event-based* scheduling mechanism for the simulator. Such a scheduler would increment its simulation clock based upon the next event to be processed. This generally implies that time increments are neither uniform nor based on a predefined fixed resolution. Based on this premise, the basic bookkeeping required by the scheduler is shown in Fig. 5.14.

Each DSL module requires queuing facilities for incoming messages, since a module processes one message at a time. This implicit queue is represented in the figure by the *message_queue* data item.

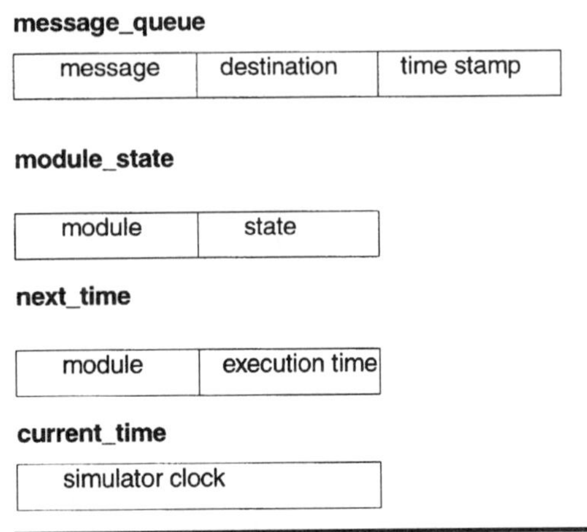

Figure 5.14 Scheduler data structures.

The item is essentially an event list and maintains an ordered list of messages to be processed on a first-come–first-served basis. A variety of data structures can be utilized to implement the *message_queue*. Some common ones are linked lists and tree and heap-based structures. An implementation based upon lists is provided in Chap. 7. The entries in the structure are the messages to be processed along with their destinations and a time stamp indicating the time at which they arrived at a module's queue. The time stamp is required to track statistics on the amount of queuing time for each message.

In addition to the list of messages waiting to be processed, the scheduler must also maintain a set of messages that are being processed. This is best explained by an example. Assume a module receives a message *m1* and immediately begins to execute its associated behavior. Assume also that this behavior requires a finite amount of simulated time to process the message (represented by a *delay* statement in its behavior). To accommodate this, the scheduler would distinguish the message from others as being *processed*. If any other message arrives at the module during this period, it must wait for the module to complete its operation on the message that is being processed. Hence the new message would be time-stamped and placed in the *message_queue*. The distinction can be managed either by using a separate data structure to keep track of messages that are being processed or by adding an extra field for the entries on the *message_queue* to indicate the time in which the message is to be processed (shown in the figure as the state entry).

Management of the *message_queue* also requires information from some other environmental data constructs. The current simulation time must be tracked somehow by the simulator. This is represented as the *current_time* construct in the figure. There also needs to be a place holder for identifying delayed modules and the time at which they are available to process further messages. In the figure this is indicated by the *next_time* construct.

The management of the *message_queue* is important, since some unwanted simulator behavior may occur. The problems occur in cases where messages may be created and scheduled continuously with no time delay. In such cases, other scheduled messages may never be processed. For example, Fig. 5.15 depicts a possible scheduling problem.

A message *m* is scheduled for processing within the current simulation time period. There are also other messages scheduled in the same time interval. Assume that the behavior of message *m* is such that it sends a message *m* with no delay. This implies that the message is to be scheduled at the next available time slot. A possibility for scheduling the message is to place it on the top of the *message_queue* as shown. However, since the message essentially produces the same message continuously, if this is done, the other messages will never be processed. A more reasonable implementation would be to place all new messages at the end of the queue so that any waiting messages are eventually serviced.

Although the message scheduler issues are important, they should be transparent to the designer, since they do not pertain to the

message_queue

message	destination	time stamp
m	source	t0
other	other	t0
...	...	...

current_time

t0

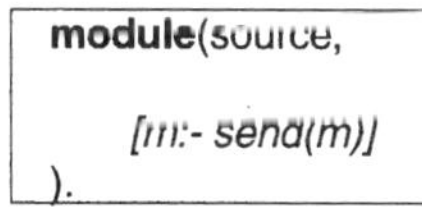

Figure 5.15 Scheduling problem.

semantics of the model. Hence these are considered implementation problems and not modeling ones.

Message resolution

One of the functions of the DSL interpreter is to execute the DSL commands, which include communication statements. These statements pose an additional problem to the interpreter. Messages can be transmitted or scheduled for transmission to unspecified modules or ports. This implies that the processor must be able to determine which module should receive the message. A resolution mechanism is required to make the correspondence from the sender to the destination module.

A means for resolving message destinations is to use an algorithm to deduce the final destination module. The approach can utilize the backtracking feature inherent in Prolog to attempt to find a module that is both connected (through *path* statements) and capable of understanding the message. An algorithm for the destination search is described as the following:

Let Source be the sender of a given message through an unknown port or destination.

Let Port be a valid output port available to sender.

0. Assign *sender* = *Source.*
1. Choose *Dest_port* and Dest such that there exists a path(*sender,* Dest, [Port, Dest_port]).
2. If Dest is a module, then

   If *message* ε *{Behavior}*

      where *{Behavior}* is the set of all behaviors within module Dest,

      then *final_destination* = Dest:

      Exit algorithm.

   else backtrack and repeat step 1 with a new Dest.
3. If Dest is a higher-order module, then

      sender = Dest, Port = Dest_port, repeat step 1.

The algorithm determines a path moving away from the source module to potential destination modules. This is accomplished in step 1. Step 2 is the halting condition, which is the discovery of a module (*final_destination*) that contains the intended message within its set of behaviors. On the other hand, if at this point the message is not within the module's behavior set, backtracking is utilized to search for alternative modules. Consequently, the search is a depth-first search of destinations.

Step 3 addresses the case in which the destination is a higher-order module. The sender and port are then assigned to the higher-order module name and port. It should be noted that there are also provisions to ensure that the algorithm always moves away from the source module (except when backtracking), so that messages do not bounce back to the source.

The algorithm returns an error condition if there is no possible destination with a connection to the sender. The environment then will attempt to search for modules that are not connected to the sender, but that contain the message within their behavior set. If a module that can accept the message is discovered, then a pseudo path must be defined by the environment. If no module is found, the message is trapped and an error message is generated.

Performance

The execution time of a simulation scenario is a concern for large models. As the number of components in a model increases, so does the number of events which must be managed by the simulator. The number of events (or messages to schedule) puts a considerable strain on the simulator performance. Therefore, techniques applicable to addressing the number or processing time of events are possible avenues for improving on the simulation time.

To address the simulation time, it is necessary to restate that a typical DSL simulator simply provides an *emulation* of concurrency. In other words, it gives the perception of simulating multiple events within the same simulated time period. In reality, the events are being scheduled in a serial fashion. The reason for the emulation is the standard single-processor Von Neumann architecture adopted by most computer platforms. Since the processor cannot handle concurrency in a strict sense, it must serialize the information and give the appearance of concurrency. Alternatively, if the computer platform were to utilize a parallel or distributed processor architecture, the concurrency inherent in DSL models would be better exploited. This, in turn, could provide faster simulation executions.

The execution of DSL models across multiple processors requires some additional creativity. The implication is that the underlying simulator would no longer execute within one processor, but within many. Issues arise regarding how simulator clocks are synchronized and how messages are scheduled and executed in a distributed context. Two types of schemes exist to address distributed simulation timing. The first are *pessimistic* or *conservative* approaches. In these schemes, the simulation clocks (or clock) are somehow kept closely synchronized between the distributed simulators. This ensures that each clock increment is executed concurrently across all simulators to maintain the causality of events. The second approach is an *optimistic* event management scheme, in which simulators process their respective messages at their own pace regardless of the state of the rest of the simulators. As a consequence, occasionally a message may be inadvertently processed before an older message. Therefore a mechanism must exist for recovering from the erroneous simulation. An example of how these can be addressed is through the use of an optimistic distributed simulation paradigm called *Time Warp,* first introduced in Jefferson (1983).

Time Warp scheduling. A distributed simulator implies that the basic scheduler or event manager of a simulator is deployed across multiple processing units and somehow permitted to operate in unison. The Time Warp scheme is one way of exploiting the power of parallel computation. In addition to local event processing and time management, each simulator requires some additional overhead.

- A history of the state of the system (set of modules) within the auspices of the simulator needs to be recorded and maintained. In particular, states of resources within modules must be traceable.
- A time-sorted output queue for messages to be sent to other simulators as well as an input queue (*message_queue*) is needed. The output queue is used to temporarily store messages destined to be sent to other simulators.
- A list of messages that were sent by the simulator is kept.

The simulators execute messages as in the single-processor case, one at a time in a time-ordered fashion. However, there is no global simulated clock. The result is that each simulator operates at the rate of its own local clock. As before, each message is time-stamped by its simulator indicating the time the message was generated.

A problem occurs if a simulator receives a message from another simulator with a time stamp (t) smaller than its local clock time (T). Since the simulator is operating at a future time relative to the message, there is a possibility that the message could have affected the state of the system if it had been processed. Hence, to maintain the causality of the system, the simulator must essentially *roll back* all its computations to the point just before the message was created.

Rollback entails going to the time within the state history indicated by the arriving message. The modules are then restored to their states as defined in the state history, the simulator clock is set to the new time, and the message is inserted into the *message_queue.* All states with a time stamp larger than the current one are discarded.

In addition, the list of messages sent must be reviewed. The messages that were sent after time *t* are no longer valid and may cause incorrect computations to occur at their destination modules. To resolve this, *antimessages* are sent to the same destinations.

Antimessages are copies of the original messages which were inadvertently sent, but possessing a flag identifying them as antimessages. The task of these messages is to pursue their counterparts and annihilate them before they can do further damage. The simulator first checks to see if any of the messages that match the antimessages are still in its output queue. If so, the corresponding message is discarded, and all is well. Otherwise, the corresponding message may be in the destination simulator's *message_queue,* where it can still be removed. The next possibility is that the message has already been processed by the simulator, and a rollback needs to be performed on the destination as well.

The Time Warp scheme has some costs associated with it. Maintaining a history of states with all local variable information can be memory-consuming. Each rollback also implies that at least one simulator performed unnecessary computations. There is also an overhead associated with destroying messages with antimessages. Studies have shown that these costs become significant in small models, but that the performance benefits of distributed simulation tend to overtake the costs for larger models.

DSL models fit nicely within the Time Warp paradigm, since communication and message management is very similar. Minimizing the number of messages transferred between simulators is an important consideration in distributed systems. A single model must be partitioned into smaller parts to be processed by each simulator. DSL higher-order modules tend to group modules that communicate often with one another and can be a basis for partitioning the overall model. This is just a guideline and is not always the case.

Preconditioning. Another mechanism for speed-up of DSL models is preconditioning of modules. This implies detection of repetitive properties in module behavior which result in excessive overhead on the simulator. Consider a module which has as one of its possible behaviors a form such as

```
loop_message:-
        delay for time period t,
        send external_message to...,
        send(loop_message).
```

This type of behavior generates a message (*external_message*) every simulated time period *t.* To generate the *external_message,* an extra message (*loop_message*) is required so that the behavior is repeated.

This latter message is termed a *looping* message. The result is that the scheduler will schedule two messages each *t* time units. If there are a significant number of modules in the model which exhibit this type of behavior, there will be an unnecessary performance deterioration in the simulator. Hence scheduling the looping messages is an unnecessary operation.

Preconditioning the DSL model for looping messages during simulation initialization can present some performance benefits. Detection of looping messages is trivial. If a module is parsed and a message contained within a *send* statement is the same as the header of the DSL behavior, then it is a looping message. Any other messages generated within the same behavior are termed the *external_messages.* Also, any delays specified within the DSL behavior description are associated with both the looping and *external_messages.* The basic data structures used in preconditioning looping messages are shown in Fig. 5.16. The steps in preconditioning are as follows:

- Determine if a module possesses a looping message. This is deduced by matching a module's behavior header to messages sent as a result of the behavior. For example, the looping message is shown in bold for the segment of behavior:

```
looping_message:- delay(t), {send external_messages}, send(looping_message).
```

message_queue

message	destination	time stamp
m1	r	T

Special_events_queue

message	source	destination	time stamp	delay
e1	s1	r1	t0	d1
e2	s2	r2	t1	d2
...	...	...	...	...
...	...	...	...	...

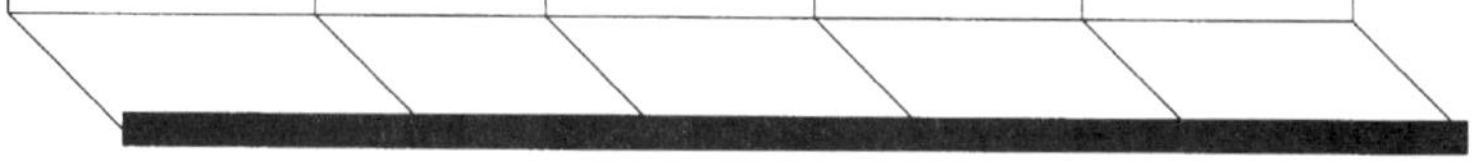

t0<t1<t2 ...

Figure 5.16 Data structures for preconditioning.

- Remove the behavior specified by the looping message. Place the *external_messages* in the *special_events_queue* (described below). Set the delays in the queue to the time parameter specified by the DSL *delay* statement.
- All initialization rules associated with the *looping_message* are now applied to the *external_messages* in the *special_events_queue.*

In addition to preconditioning the modules, the basic *message_queue* concept must be expanded for the scheduler. Assume the existence of two message queues:

- As before, *message_queue* maintains a time-ordered set of messages to be processed by the simulator.
- A second time-ordered queue (*special_events_queue*) maintains a queue of *external_messages.* For each entry, the queue records the message, source and destination, time stamp, and delay. These parameters are derived directly from the message's associated looping message. These are stored in the queue in the order of their instantiation.

The basic operation of the revised scheduler is as follows:

- The scheduler checks the time stamps on the top entries of the *message_queue* and *special_events_queue.* The message with the smallest time gets processed.
- If this message happens to be in the *message_queue,* it is processed by the scheduler in the same manner as before.
- If the message is in the *special_events_queue,* this implies that a (set of) message(s) is to be processed as a result of a looping message. The message(s) is processed, and any resulting events are scheduled into the *message_queue.*
- At the completion, the time stamp of the message is updated to equal the sum of the *delay* parameter from the *special_events_queue* and the current simulation time.
- The message is then reinserted at the end of the *special_events_queue.*

The Environmental Model

This chapter has focused on issues of design execution. No distinction was made between the environment model and the system under design. The rationale behind this is that both models are essentially DSL representations and require the same design execution support.

However, the environment model can be viewed as a specific case or subset of the general applicable model space for DSL modules. Two classes of modules may be established to represent the type of models needed to define an environment's behavior. These classes were defined as source generator and probe models. Library modules created and classified within these two classes can be used over and over to capture the specific needs of the surrounding environment.

Source generators

Source generators consist of traffic generators, random-number generators, trace messages, or any other type of input that needs to be produced for the system under design. Most of these are reusable components. For example, random-number generators are obvious candidates to be used in many different designs. Therefore, a library consisting of a good selection of generic random-number generators is covetable. A list of some of the common source generators is as follows:

- *Random-number generators.* A slew of random-number generators are possible. These are based upon a mathematical distribution such as the *exponential* or *erlang-k*. A library of these sources is a basic necessity for modelers.
- *Error generators.* These sources work in conjunction with other sources to inject failure or error messages into a system. They are useful for verifying robustness and system reliability.
- *Discrete clocks.* Clocks producing a message at programmable periods are invaluable in creating controlled timing and traffic loads. A timer and clock description was presented, but many other types of clock behavior can be utilized.
- Application-specific traffic. Some designs may require specific data streams as input. For example, a sample stream of voice or video traffic may be needed for a multimedia system design. In these cases, it may be useful to maintain a library of such sources.

Probes

Probes are the means for observing the intricacies of the model during design execution. They are receptacles for messages generated by the system under design. The following is a list of some common probes:

- *Counters.* A useful set of modules are those that monitor the number of given events in the system. The simulator should perform basic counts of messages arriving and leaving module queues as part of its statistics gathering. However other count-monitoring modules may be required to detect patterns at other points in the model (such as total number of messages with a given parameter).

- *Triggers.* Modules may also generate some type of feedback to indicate the occurrence of an event. The feedback can be in the form of a visual indicator (such as animation or graphics) or a message to activate a source.
- *Error traps.* Errors which occur during design execution can be handled by the simulator or by specific probes. Designing probes to handle specific error conditions customizes the treatment of error exceptions, but it is unsuitable for the general case.
- Application-specific probes. Just as in the case of library configuration for generic modules, probe modules can also be defined to capture specific application information. For example, a probe may be configured to interpret a sequence of video messages to generate an image on a simulated or real screen. In this case, a probe behaves as a monitor or receiver.

Chapter 6

Design Synthesis

Until this chapter, we have focused on issues related to the modeling of systems to be designed. The focus has been on the construction and support of the models, their execution and analysis. These models have been assumed to be represented in a DSL format. After the design execution activities have been completed, further support is needed to transform the DSL representation into a lower-level representation.

The term *design synthesis* is used in this chapter to encompass all the remaining activities required to reach the final objective of the architectural-design process: to generate a lower-level description of the design (the *target model*) suitable for use by existing design aids. In a strict sense, some of the activities within design execution can be viewed as synthesis activities as well. For example, the progressive refinement of DSL models by the user and environment constitutes a synthesis activity. Furthermore, the term may have different connotations at different levels of abstraction. For the sake of clarity, it is best to provide a definition of the synthesis terms that are used in this chapter.

> DEFINITION: *Synthesis* is a term used to denote a process which results in the transformation of an abstract (intermediate) model into a more detailed (target) model. That is to say, a more complex model is formed from simpler one(s).

Figure 6.1 illustrates the basic synthesis process. In this example, an intermediate model m is transformed into a target model m'. The transformation results in a new interconnection scheme, resulting in (a possible) new composite behavior.

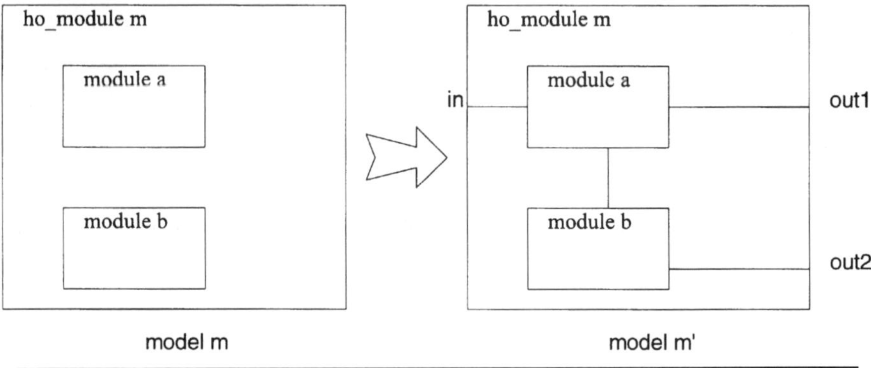

Figure 6.1 Synthesis process.

DEFINITION: *Design synthesis* denotes all the synthesis activities after the design execution phase that are required in order to transform an (intermediate) model into a lower-level (target) design abstraction.

Figure 6.2 shows an example of design synthesis. The example illustrates the transformation of an intermediate model to a target model through successive refinements to the original model. The refinements enable the model to be transformed into a less abstract form. The design synthesis activities constitute the transformation of the final refinement of the intermediate model m_{final} to the target model m''.

This chapter describes the outstanding issues remaining in the refinement of the translation activities that will enable design synthesis.

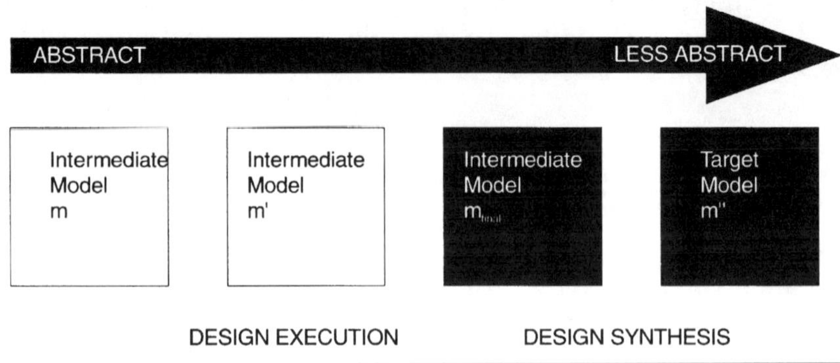

Figure 6.2 Design synthesis.

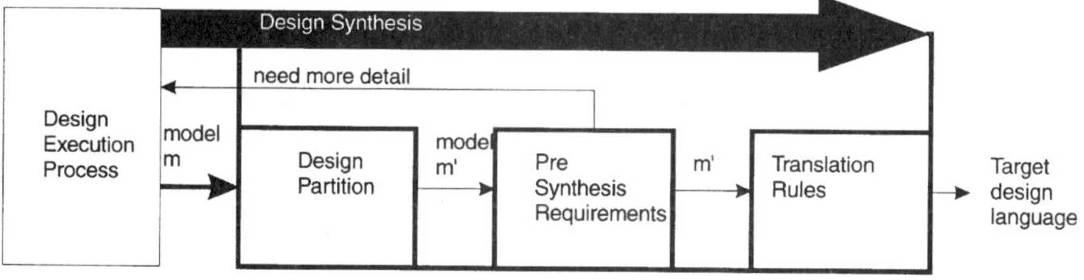

Figure 6.3 Details of design synthesis process.

Synthesis Process

From the perspective of architectural design, synthesis is achieved when the intermediate representation can be translated into a form usable by existing design tools which will subsequently result in the eventual fabrication of the design. Figure 6.3 depicts the design synthesis process as part of a set of activities involving design execution, partitioning, and translation rules. When the designer is satisfied with the simulation results, the model's synthesis may be completed with a translation of the model to a lower-level description.

Design partitioning is an activity which follows the design execution phase. Ambiguity within the intermediate model must be resolved by the designer and environment during the design execution phase until a synthesizable design is reached. The design can then be partitioned to reflect software and hardware components of the system. The partitioned design (represented as m' in the figure) must undergo some tests to ensure that it can be synthesized to the target language.

At this point, when synthesis to a lower level is possible, a mapping must take place from low timing detail to higher timing detail. To accommodate this, the environment must be able to fill the essential details through language-dependent translation rules. Some type of compromise is required by the synthesis tool (such as cost, area, or performance) so that details can be added. It is this fact that can make synthesis tools very poor performers as compared with their experienced human counterparts.

Target Languages

The target languages are representations that can be utilized by other design environments to move the DSL model into an implementation mode. DSL can generate a software or hardware representation of elements of the model. Consequently, the target languages are those applicable to either of the two disciplines.

Hardware description languages

Standardization of design languages has been successful in the hardware arena. Currently two dominant hardware description languages are used in the industry: VHDL and Verilog. The former is an earlier standard issued by the IEEE, and the latter is a newer standard based on the Verilog language. Both are simulation-oriented languages and incorporate simulator-directed constructs within their semantics. For example, a designer using VHDL must have a basic understanding of how the simulator will execute the model. Translation to both languages is possible, with the general procedures being similar. This chapter will cover only translation of DSL to VHDL as an illustration of the steps followed in the process.

Software descriptions

Software synthesis is a prospering field; however, it still lacks some of the fundamental achievements seen in the hardware domain. For example, software construction is still not as well disciplined as hardware construction. Most modern hardware components that are used in modeling libraries are well defined and follow accepted standards and notation. Software libraries do not follow a globally accepted form. Also, there are a significant number of possible target languages, such as C and ADA. Unfortunately, the representational forms of coding tools that can accept high-level software models are not necessarily compatible with one another. Hence DSL model translation to the descriptive representation of one of these languages will be different with each tool.

Despite the lack of a standard representational form for software, all is not as bleak as it appears. Unlike hardware, software models are not constrained by subtle timing dependencies and are more readily convertible from one paradigm to the next. This is to suggest that relationships captured within a given paradigm can be expressed in other forms. Hence, if a general procedure for the translation of DSL to software is provided, it may be applicable to a large set of the tools.

One interpretation of DSL modules is to view them as communicating finite state machines. This is depicted in Fig. 6.1, where the circles represent states of a module (a circle surrounded by a square represents a starting state). Messages are used to instigate changes in state, and resources maintain the state information. This view can be applicable to many software representation schemes and computer-aided software environments.

Design bounds on the software can also be established. The Petri-net formalism as applied to modules can be utilized to transform the DSL representation into a representation suitable for coding. A trans-

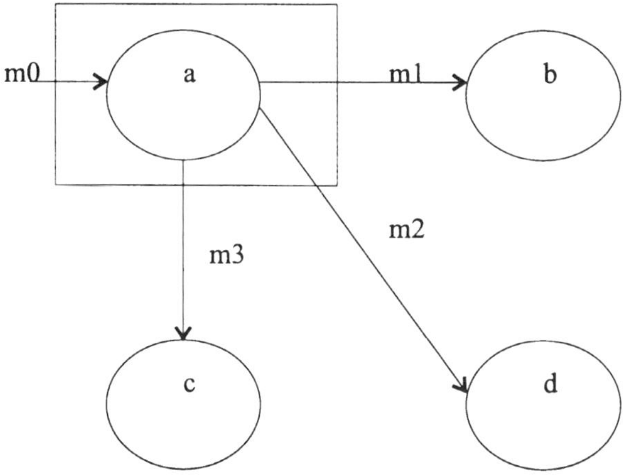

Figure 6.4 Communicating FSM representation of modules.

form theory for defining software performance using Petri-nets was presented in Lewis (1982) and could readily be applied in this instance. In this reference, software is represented as Petri-nets with time attributes attached to transitions. The transitions represent delays in executing a specified segment of code, analogous to the delay statement in DSL. An example of a representation of an *if-then-else* statement is shown in Fig. 6.5. The delays associated with the execution of the statement are given by *t1, t2,* and *t3.* A probability *p* represents the probability of the condition (*cond*) being true. The interpretation of this in terms of a Prolog specification is also straightforward. A DSL description of this could be the following code fragment:

```
(b(Cond):- Cond,delay(t2)),
(b(Cond):- not(Cond), delay(t3)).
```

The message name is irrelevant in this example. *Cond* is assumed to be a logical variable that is being tested. The time to test the condition (*t1*) can be interpreted as the queuing time for the message.

Through a set of transformation and decomposition rules, the Petri-net is collapsed into a single place and transition. A probabilistic formula attached to the transition reflects the average delay parameter of the transition. For example, the *if-then-else* statement can be reduced to the structure shown in Fig. 6.6. In the DSL context, the formula represents an average delay associated with a module based upon the probability of executing its different behaviors. With the premise that each module's behavior can be approximated by the per-

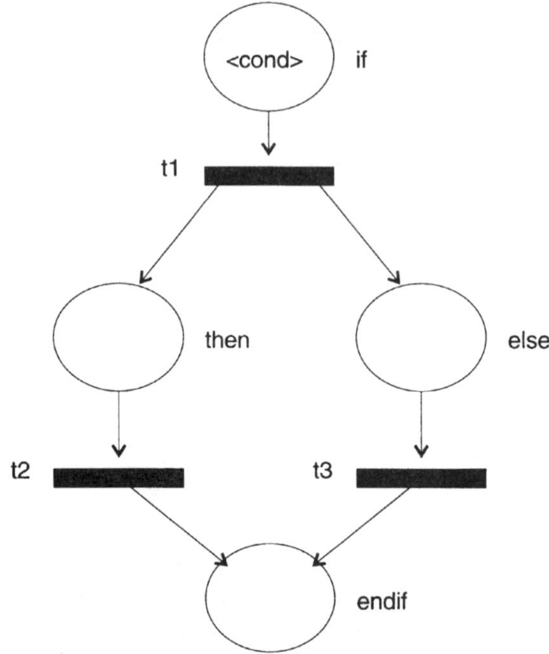

if cond=true then {delay t2} else {delay t3} endif

Figure 6.5 Petri-net representation of if–then–else statement.

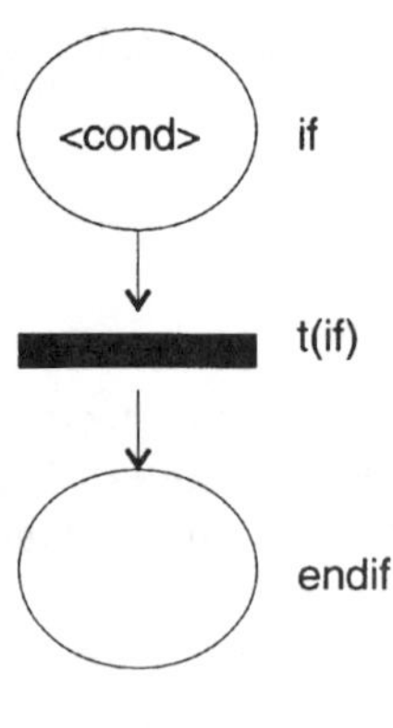

t(if) = t1 + p.t2 + (1-p).t3

if cond=true then {delay t2} else {delay t3} endif

Figure 6.6 Reduction of Petri-net.

formance bounds depicted with a single place and transition, a net can evolve for the larger DSL module. This net permits the estimation of performance bounds on the software through the average delays tagged to the transitions. The reader is encouraged to review the reference for further details.

In the telecommunication domain there are also a host of other formalisms that can be used, such as SDL or LOTOS. The process for translating DSL to the respective software form can follow an approach similar to that described below for hardware. The distinguishing features of the translation are more in the syntactical details of the target language.

VHDL overview

Since VHDL will be used as a target language, an overview of its constructs is presented for readers unfamiliar with the language. This overview will describe the elements of VHDL as it pertains to the translation of DSL primitives. Hence, only a subset of the VHDL language and modeling styles will be addressed. As an experienced VHDL designer will attest, the language provides multiple ways in which behavior can be described. However, only the concurrent process description view is introduced here. The reader is referred to the VHDL standard for a complete description of the language. All bold characters in the following VHDL descriptions identify reserved words in the language.

VHDL components can be defined using an *entity* construct. An entity is a combination of a declaration and an architecture body description. The declaration provides a description of the interface of the component (a type of black-box view). The interface, as perceived by other components, is a set of ports. These points of connection permit data flow from one component's port to another. The direction of the port is identified within the port definition with the reserved *in, out,* and *inout* statements. Each port is declared as a specific VHDL data type. An example of an entity declaration for a timer is

```
entity Timer is
      port(
      Reset: in Bit;
      Output: out Bit);
end Timer;
```

While ports define the points of communication in a component, *signals* actually connect the components through the ports. A signal is used to transmit information between entities and comprises a value. This value belongs to a certain predefined set of VHDL types. A signal may change its value through signal assignment statements (discussed below).

The second portion of an entity is the architecture body description. This provides a representation for the behavior of the component. The architecture body may contain a set of local declarations for signals and variables and also a set of *processes*. The VHDL *process* is a powerful construct for capturing concurrent activities within an entity. Concurrent processes can be described by statements bounded by a *begin* and *end*. Hence the architecture body for a timer example resembles the following:

```
architecture Timer_behavior of Timer is
begin
{VHDL behavior statements}
end Timer_behavior;
```

A process is activated or executed when there is a change in the state of a signal contained within the process. An execution consists of the systematic execution of the VHDL behavior statements. These may consist of an assortment of language primitives such as assignment and control statements. Perhaps the most utilized one is the signal assignment. The form of the statement is

```
signal_name< = value after delay;
```

The *signal_name* is a signal which is assigned a new value after a specified delay. The delay is a time expression. While signals can be used for communication, local variables can be used for maintaining information relevant to an entity or storing temporary information.

Other useful VHDL behavior statements are those permitting control flow. Conditional statements similar to those in many standard procedural languages are possible in VHDL. An if-then-else–type construct can be used to control program flow.

A host of different data types are available to the user. Apart from standard built-ins such as *bit* and *integer*, the language permits enumerated types. This allows programmers to define a set of literals belonging to a new type. For example,

```
type traffic_light is (`red', `green', `yellow')
```

declares a new type. Signals are bound to the type *traffic_light* and may assume one of the three identified values.

The complete description of the timer example is given in Fig. 6.7.

```
entity Timer is
    port (
    Reset, Clck: in Bit ;
    Output : out Bit) ;
end Timer ;
architecture Timer_behavior of Timer is
begin
variable Value : Integer := 0;
    process
    begin
        if (Modulo_limit > Value) then
            Value := Value +1;
        else
            Value := 0;
            Output <= '1';
        endif;
    end process;
    process (Reset)
    begin
        value:=0;
    end process
end Timer_behavior;
```

Figure 6.7 VHDL description of timer.

DSL–VHDL differences

It is important to appreciate the differences between the intermediate and the target language. A language comparison would reveal some common features and some significant differences. The latter are a reflection of the intended application of the languages. As one moves from an abstract level of design (architectural) to more detailed levels (such as RTL), there is a strong shift of detail in the models created at each level. At the architectural level, little timing detail is provided compared to the lower levels. However, freedom is given to structural aspects of the model so that design exploration can be facilitated. At this point of design, a certain level of ambiguity is introduced, which is a consequence of the flexibility in modeling with loosely typed constructs.

The strong distinction between design exploration (a key requirement at the architectural level) and design formalization (a major requirement between the lower levels of abstraction) dictates the need for different languages. Hence, DSL is an internal language to support the former, and VHDL the latter.

The major difference in the languages is the level of abstraction in which they are to be applied. VHDL's applicability is approximately limited to RTL to circuit-level design. Application to higher levels is also possible, but it is unlikely for several reasons. The language contains many simulator-oriented semantics (such as wait on event of a simulator), which are superfluous at higher levels of design. The language carries primitives which are not highly desirable for architectural-level designers (especially not for software designers), such as signals and transport (low-level timing) commands. Hence architectural-level representation would be possible, but it would be highly cumbersome, since the language was not intended for this purpose.

VHDL is a good language for hardware formalization, but very poor for design exploration. This is not a negative feature, but is again related to the area of applicability of VHDL. As an RTL description language, it provides strong typing of communication and timing between entities. However, as the abstraction level increases, detail is lost and fewer restrictions on the typing of the entities are required to permit design exploration. This necessitates the use of a language such as DSL.

Both DSL and VHDL use a simulator for timing verification. However, simulator semantics are not embedded within the DSL language, as they are in VHDL. Differentiating the simulation semantics from the modeling language is a highly desirable feature for an architectural language, since it permits formal verification. Such verification techniques have been cumbersome and fleeting for VHDL representations, whereas Petri-net-based verification (and other types) can be readily applied to DSL modules.

Partitioning

During the design of models for the system under consideration, DSL makes no distinction between software and hardware modules. However, the language has provisions to allow the designer to identify these modules. This is of significance for hardware synthesis, since software modules will not be synthesized to the target hardware language, VHDL. A module can be identified as being either a hardware or software entity simply within the predicate *module_type:*

```
module_type_definition:: = module_type(dsl_name,synth_type).
synth_type:: = software
```

```
|hardware
|system
|protocol
```

In this description, *dsl_name* is the name of a module. Apart from the obvious labels denoting hardware and software elements, there are also identifiers for *system* and *protocol.* The *system* identifier implies that the module is a supporting module (such as an environment module) and is not synthesized. The *protocol* argument indicates that the module is used to implement a protocol between other modules and can be synthesized into a VHDL procedure or function.

If the *module_type* predicate is not defined by the user, the respective module is assumed (by default) to be a *system* type and is not considered for synthesis.

The representation mechanism provided by DSL facilitates the description of generic modules. These modules are defined entirely in terms of their potential behavior with respect to the environment and are free to bind to software or hardware constructs. Since there is no default notion of hardware and software, co-design modeling can be facilitated by DSL.

Co-design concepts within DSL are illustrated in Fig. 6.8. Modules can represent behavior and structure (HO-modules), so that many different abstraction levels of hardware can be described. Software behavior can similarly be captured by the behavior of a module. Since each module

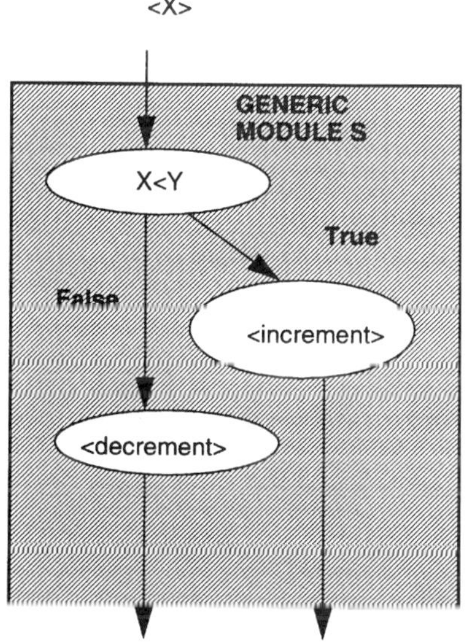

Figure 6.8 Software description example *(1995 IEEE).*

may execute one particular behavior at a time, it is straightforward to describe serial program execution in terms of delays and program control. This is depicted within the figure by the shaded generic module *S* (for software). The ovals graphically represent a particular behavior, while the arrows indicate the program flow control. The names within "<>" indicate a name of a behavior defined within the module. In this simple example, module *S* will accept a message with parameter *X* and either increment or decrement the value, depending on the condition $X<Y$ (where *Y* is assumed to represent a local variable).

Concurrent software and hardware can also be described with a set of modules, confined within a higher-order module. An example of this is given in Fig. 6.9. Higher-level program control is described by the module interconnects, while local module control flow is described by the behavior of each module.

In this example, a module *PROC(X)* spawns three identical messages, indicated by the *fork* behavior. This behavior would typically be described in DSL in the form

```
fork(X):-send([proc(1),proc(2),proc(3)],_, X).
```

The three concurrent processors (hardware or software) inherit a part of their behavior from the generic module *S*. Consequently, each

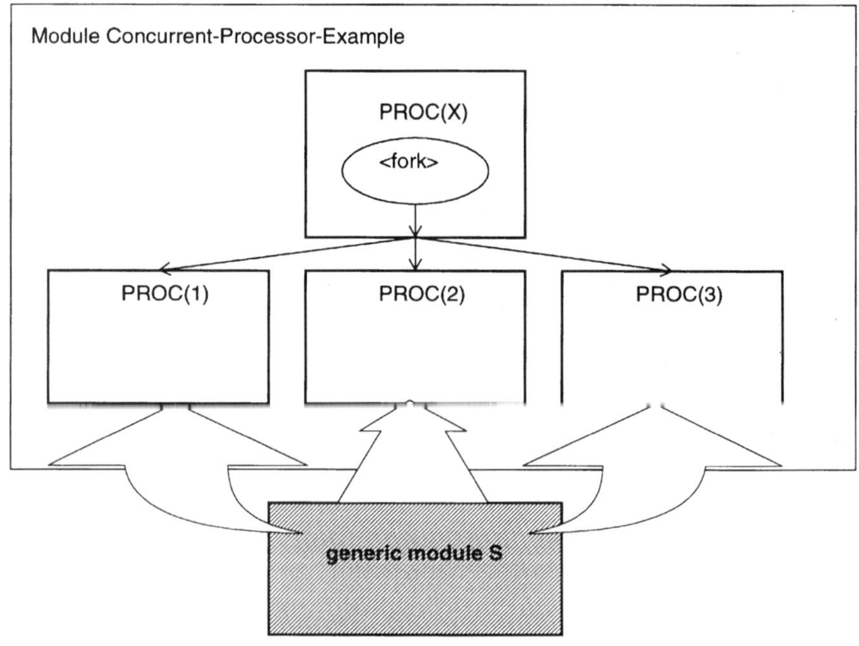

Figure 6.9 Fork example *(1995 IEEE)*.

processor will process the message *X* concurrently. It should be noted that the processors *PROC(1–3)* can represent either hardware processors or software processes. Inheritance is the typical way in which a hardware module may absorb the behavior of a software module.

If a library with generic software modules representing specific functions such as branching, looping, and sorting is maintained, then these modules can be inherited by other modules as required. Parameterized generic modules allow for the dynamic definition of module resources and delays upon invocation. This facilitates the description of reusable software modules such as *sorter,* which may accept messages such as *quick-sort(SIZE)* and *binary-sort(SIZE).* The parameter *SIZE* may be used by the generic *sorter* module to determine the delay of the particular sort algorithm.

Presynthesis Support

During the course of design exploration and simulation, the abstract DSL model may be refined to a more detailed structure. For example, ports may be identified and created on demand by the environment. A DSL model is not always favorable for translation to another paradigm. It may need further refinement before it can complete the synthesis process. The point in time at which a DSL model can be translated to an equivalent VHDL counterpart is determined by a set of presynthesis requirements. The following are the minimum set of requirements that must be met by the DSL model.

Requirement 1

All undefined communication paths within an initial DSL model must be resolved. This implies that all undefined ports within *send* statements have been instantiated by the environment. The port labeling convention used is dependent upon the particular one an environment adopts. The requirement ensures the existence of a physical path between communicating modules so that an equivalent structure can be constructed in VHDL.

Requirement 2

All message types must have been resolved for every port. This essentially means that all messages transmitted by *send* commands have been identified to their respective ports. The synthesis process requires this information to correlate the ports with equivalent VHDL signal types.

Requirement 3

The environment must have defined all destination modules indicated by DSL *send* statements. The environment can deduce most connections from the *path* statements. However, outstanding connections must be established by the environment or the user.

Requirement 4

Multiple connections from output ports must be resolved. The implication is that DSL *send* statements with multiple destinations of the form *send([b,c,d],aout,message)* must be handled by the environment. The issue during synthesis with such communication is that the output port is used to transmit directed messages (messages for which the destination is given). This implies that a separate signal or a bus protocol (utilizing the destination as part of the address resolution) must be used in VHDL to realize the communication.

Requirement 5

Multiple connections to input ports must be resolved. This requirement can be satisfied in two ways. The first is to ensure a unique input port name for each *path* statement. The second is to introduce a *protocol module.* This module is conceptually different from normal modules. It does not get synthesized to a VHDL entity or process, but defines communication protocols between modules. This is a natural step when modeling system-utilizing protocols for bus arbitration or shared mediums. A module is tagged as a protocol module simply within the predicate: *module_type(Module_Name, protocol).*

DSL-to-VHDL Translation

The parser maintains a set of default variables and conventions during synthesis. It also synthesizes to a subset of VHDL, since commercial VHDL synthesizers cannot handle the full VHDL language. The applicable subset of VHDL will vary with each synthesis tool. The corresponding VHDL statements presented in this section are generally acceptable in most cases; however, some tools may not support a particular form. In this case, additional attention (such as further translations) may be needed before these statements can be used with the particular tools.

Translation rules

The following is the set of rules for translating DSL models into VHDL. A summary of the rules is presented in Table 6.1 for reference.

TABLE 6.1 Summary of Translation Rules

Id	DSL statement	VHDL statement
Declaration	outport_list(Module, OPort(o1,…,om), [Behavior(a11,…,a1n),…, Behavior(aj1,…,a1ij)]). inport_list(Module, IPort, [Behavior(a11,…,a1n),…, Behavior(aj1,…,a1ij)]) module(Module(p1,…,pl), [{behavior}]).	/* within declaration of Module: OPort_o1…_om_type is ('Behavior(a11,…,a1n)',…, 'Behavior(aj1,…,a1ij)'); IPort_o1…_om_type is ('Behavior(a11,…,a1n)',…, 'Behavior(aj1,…,a1ij)'); signal OPort_o1…_om: OPort_o1…_om_type; signal IPort_o1…_om: IPort_o1…_om_type; Module_p1…_pl: process(
Behavior	(Behavior(a1,…,an) :-{conditionals and action})	if in 1 = behavior(a1,…,an) then {conditionals and actions} endif;
Conditional	check_res(Resource(r1,…,rz),(v1,…,vy)), {actions}.	type Resource_type is record v1 : VALUE1; v2 : VALUE2; … vy : VALUEy; end record; variable Resource_r1…_rz : Resource_type; if Resource_r1…_rz = (v1,…,vy) then {actions} endif;
Action	delay(DELAY)	used during translation of DSL send statement below
Action	set_res(Resource(r1,…,rz),(v1,…,vy))	Resource_r1…_rz: = (v1,…,vy);
Action	create(Resource(s1,…,sz),(v1,…,vy))	not permitted
Action	remove(Resource(s1,…,sz),(v1,…,vy))	not permitted
Action	send(Dest(d1,…,dn), OPort(o1,…,om), Behavior(aj1,…,aij))	OPort_o1…_om< = Message (aj1,…,aij) after DELAY;
Action	send([Dest(d1),…,Dest(dn)], OPort(o1,…,om), Behavior(aj1,…,aij))	OPort_o1…_om_Dest_d1< = Message(aj1,…,aij) after DELAY; … OPort_o1…_om_Dest_dn< = Message(aj1,…,aij) after DELAY;

These rules utilize some labeling conventions which do not reflect upon the translation process, but constitute a good example of the typical issues that will be encountered.

The following are some shorthand notations used in the description of the rules:

- A predicate *outport_list(Module, OUTPort, [Behavior])* is assumed to be created by the environment. This identifies all messages that the *Module*'s output port *OUTPort* will support. The information can be extracted from the *send* statements within the module and appended to the *behavior* list.

- A predicate *inport_list(Module, INPort, [Behavior])* is used to identify all the messages that can be received through a *Module*'s input port *INPort*. This is obtained through a comparison of the possible sources of messages to a module (found by an examination of the *path* statement) and the corresponding messages that can potentially be delivered to the respective port (this information is already available from the *outport_list*).

Rule 1. The *outport_list* or whatever other means is used by the modeling environment to record messages departing from a given port corresponds to the signal type declaration of VHDL. Hence, the DSL code of the form

```
outport_list(Module, OPort(o1,…,om),[Behavior(a11,…,a1n),…,Behavior
(aj1,…,a1ij)]).
```

is used to declare

```
OPort_o1…_om_type is ('Behavior(a11,…,a1n)',…, 'Behavior
(aj1,…,a1ij)');
signal OPort_o1…_om: OPort_o1…_om_type;
```

Rule 2. Like the *outport_list,* the *inport_list* is used as a VHDL type declaration. Hence, the DSL code

```
inport_list(Module,IPort,[Behavior(a11,…,a1n),…, Behavior
(aj1,…,a1ij)])
```

is transformed to

```
IPort_o1…_om_type is ('Behavior(a11,…,a1n)',…, 'Behavior
( aj1,…,a1ij)');
signal IPort_o1…_om: IPort_o1…_om_type;
```

Rule 3. The name of a DSL module is translated into a label for a VHDL process. Module attributes are attached to the label to create a unique name. The DSL form is

```
module(Module(p1,…,pl), [{behavior}]).
```

The corresponding VHDL form is

```
Module_p1…_pl: process(
```

Rule 4. A DSL behavior header is interpreted as a condition statement. A structural correspondence is established in VHDL. So the DSL statement

```
(Behavior(a1,…an):-{conditionals and action})
```

is translated into the form

```
if in1 = behavior(a1,…an) then {conditionals and actions} endif;
```

Rule 5. The DSL resource construct is recorded as a VHDL record type. A DSL resource of the form

```
Resource(r1,…,rz),(v1,…,vy)
```

corresponds to a VHDL type of the form

```
type Resource_type is record
v1: VALUE1;
v2: VALUE2;
…
vy: VALUEy;
end record;
```

The *VALUE1* to *VALUEy* declarations are types associated with each of the resource value arguments. The types can be deduced through a parser (shown in the next chapter). In addition, the VHDL statement

```
variable Resource_r1…_rz: Resource_type;
```

is needed to reference the actual resource name.

Rule 6. DSL resource check statements are simply control constructs in VHDL. Hence the statement

```
check_res(Resource(r1,…,rz),(v1,…,vy)), {actions}.
```

is simply inferred as

```
if Resource_r1…_rz = (v1,…,vy) then {actions} endif;
```

Rule 7. The resource assignment statement in DSL,

```
set res(Resource(r1,…,rz),(v1,…,vy)
```

transforms to a variable assignment in VHDL:

```
Resource_r1…_rz: = (v1,…,vy);
```

Rule 8. The DSL *delay* statement is translated into a VHDL time expression. Since time expressions are used in conjunction with signal assignments, the delay parameter of

```
delay(DELAY)
```

is applied to any of the following communication statements as shown in rule 10.

Rule 9. Certain DSL statements are not translatable to VHDL. These pertain to dynamic generation and destruction of resources. Hence statements of the form

```
create(Resource(s1,…,sz),(v1,…,vy))
```

and

```
remove(Resource(s1,…,sz),(v1,…,vy))
```

are not permitted within a DSL model that is to be synthesized. A possible solution to DSL behavior expressed in this manner is to mark the component as a *software* module, where it can be translated into a dynamic data structure.

Rule 10. The communication primitives in DSL are translated into VHDL signal assignments. As was described, there are different variations for the DSL *send* statements. The statement of the form

```
send(Dest(d1,…,dn), OPort(o1,…,om), Message(aj1,…,aij))
```

is described as

```
OPort_o1…_om<=Message(aj1,…,aij) after DELAY;
```

The *DELAY* time expression is derived directly from a preceding *delay* statement. In the absence of a DSL *delay* statement, the signal assignment is written without any delay.

Rule 11. A second variation of the *send* statement utilizes multiple destinations:

```
send([Dest(d1), …,Dest(dn)], OPort(o1,…,om), Behavior(aj1,…,aij)).
```

The corresponding signal assignment in VHDL requires as many signals as there are DSL destinations in the *send* statement. This is obtained by prefixing the destination name to the signal name, to obtain

```
OPort_o1..._om_Dest_d1< = Message(aj1,...,aij) after DELAY;
...
OPort_o1..._om_Dest_dn< = Message(aj1,...,aij) after DELAY;
```

Modes of translation

The translation of DSL to VHDL representations is not straightforward and can proceed in two different modes, depending upon the intended description. The particular mode of synthesis is maintained within the *mode(Module, Type)* predicate, where the *Type* variable is either a 1 or a 2, depending on the desired synthesis mode for *Module,* as described below.

Mode 1. By default, the environment can assume that causality of messages arriving to modules is satisfied by the DSL design. Most synchronous DSL designs will satisfy this assumption. In such a case, translation from DSL to VHDL follows direct rules, under which DSL ports and messages translate to VHDL signals and values, respectively. Each DSL module corresponds to a VHDL process, and DSL behaviors map to VHDL statements within a process. State or variable values and types are also defined through the DSL environment. The basic algorithm for this mode of translation is given below:

```
step 1. For each DSL higher-order module:
        begin h-o-module,
               create a VHDL entity header and port definition,
step 2. For each DSL module:
        begin module,
step 2.1.      create a VHDL process (named after the DSL module),
step 2.2.      create the VHDL process sensitivity list,
step 2.3.      translate each DSL behavior to corresponding VHDL
               behavior,
              end module,
        end h-o-module
```

The details of the algorithm are presented below for each of the steps.

- Step 1 constitutes the entity declaration. The higher-order module name encapsulating the module(s) is used as the entity name. VHDL port declarations are defined through the DSL *path* statements.
- Step 2 constitutes most of the behavior translation activities. The first substep (2.1) creates a VHDL process with a label corresponding to the DSL module name. This is a straightforward procedure, shown in Table 6.1.
- Step 2.2 establishes a sensitivity list for the corresponding VHDL process. The list is defined through the *inport_list* predicate. Every input port of a module corresponds to a signal within VHDL.

- The final step (2.3) is a translation of DSL actions to VHDL statements. Table 6.1 summarizes all the corresponding translation rules employed by the parser.

Mode 2. For asynchronous design components, the user may direct the environment by setting a predicate within the simulator called *mode(MODE)*. *MODE* establishes the mode to use during parsing. This can be 1 or 2. The latter indicates the desire to synthesize an asynchronous design. This has the effect of indicating that the implied causality and ordering of DSL messages should be maintained in the VHDL model. In such a case, the corresponding VHDL model will include an input buffer and control for each entity so that the order of arriving signal values (which correspond to DSL messages) is maintained. Each signal value within the buffer is processed in turn by the VHDL entity. Such designs may be costly in terms of hardware, but VHDL postsynthesizers can optimize them further.

The algorithm for the second mode of translation is slightly different, so as to ensure the desired timing requirements. The first step of the algorithm is identical to that described for mode 1 synthesis. The differences are within step 2. The algorithm for step 2 is:

```
step 2. For each DSL module:
          begin module,
step 2.1.  create a VHDL process "MODULE" named after the DSL mod-
           ule,
step 2.2.  create a VHDL process "queue_MODULE"
step 2.3.  create the VHDL process sensitivity list for "queue_
           MODULE",
step 2.4.  sensitivity list for process MODULE is: signal_MODULE,
step 2.5.  create FIFO queue code for queue_MODULE, and MODULE.
step 2.6.  translate each DSL behavior to corresponding VHDL
           behavior for MODULE,
        end module.
```

Essentially what is obtained by the algorithm is two VHDL processes per DSL module (shown in Fig. 6.10). One process is sensitive to incoming signals (as is the case for mode 1 processes); however, its behavior represents a first-in–first-out (FIFO) queue. Hence any new signal values are stored in the queue structure represented within the process. A signal named *signal_MODULE* is used to communicate with the second process. Another signal (*ready*) is an input from the *MODULE* process. When the *ready* signal is asserted, the *queue_MODULE* places the next value of *signal_MODULE* from its FIFO.

The *MODULE* process is sensitive only to the *signal_MODULE* signal. When a change in this signal is detected, the value is obtained and treated within the body of the process, as for mode 1 translation.

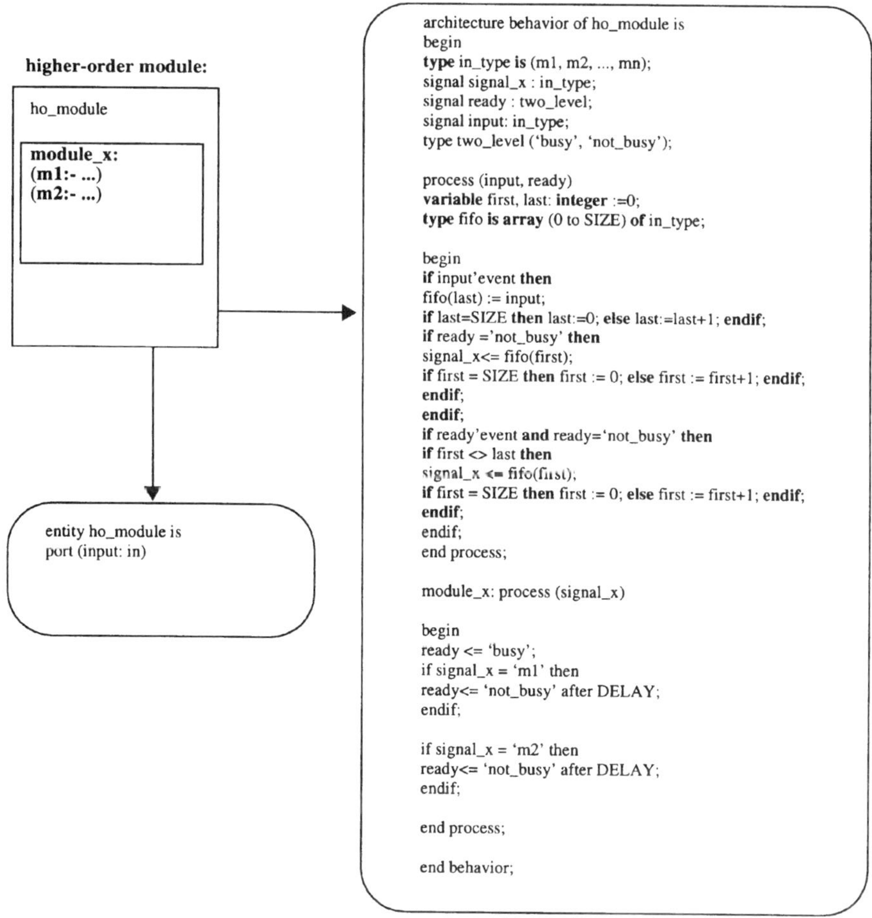

Figure 6.10 Mode 2 translation example.

At the same time, the *ready* signal is not asserted, indicating that the process *MODULE* is busy processing a message. The signal is asserted again to obtain the next message (if any).

The combination of the two signals ensures that incoming messages are correctly ordered for the process *MODULE*. Figure 6.11 presents the typical timing waveform encountered for the two processes. The first two waveforms represent the *ready* and *signal_MODULE* signals (in this example, the *MODULE* postfix is *a*). The *signal_a* is an enumerated type (*m1, m2, m3, m4* in this example). The third waveform indicates the arrival of new signal values detected at the *queue_a* process. The *ready* waveform indicates the order in which the incoming signal values (or DSL messages) have been processed.

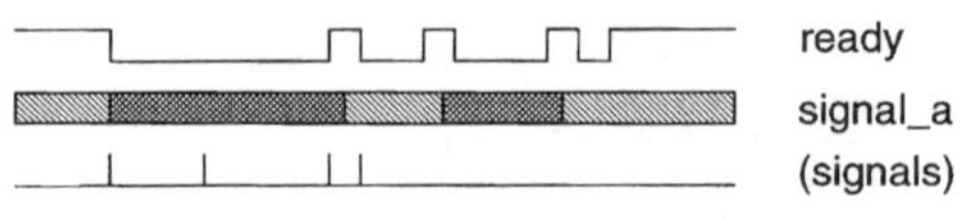

Figure 6.11 Timing waveform for mode 2 example.

The amount of attention given to timing considerations within DSL can significantly affect the eventual VHDL code. If possible, the designer should incorporate as much of the timing relationship as possible within the DSL model (especially for reusable generic modules). This will tend to produce less overhead within the synthesized VHDL code. An example is given below to help clarify the modeling issues.

Consider the example presented in Fig. 6.12. A DSL module *repeater* is presented, with one input port *input* and one output port *output* specified. The behavior of the module is simple: It sends any message arriving at its input port to the output port after a delay of 3 time units. A stream of incoming messages is assumed to be sent to the module, as shown in Fig. 6.12*a*.

The first example (indicated in Fig. 6.12*b*) shows a synthesis of the DSL module in mode 1. The resultant VHDL code does not produce the exact same timing characteristics for the input message stream.

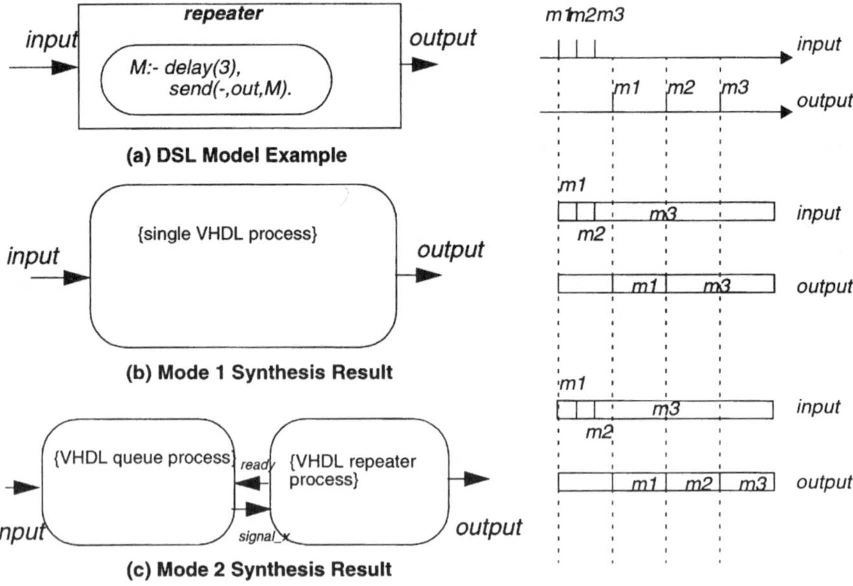

Figure 6.12 Repeater example.

The problem is that the *m3* value overwrites the *m2* value before the VHDL process can detect it. Figure 6.12*c* depicts the second mode of synthesis, ensuring the correct ordering of messages. However, it can be observed that the overhead within the VHDL model is significant compared to the first mode. Further optimization could be attempted at this point with lower-level synthesis tools. The other alternative to guaranteeing the same timing characteristics between the two representations is to enforce the timing discipline within the DSL model. In this example, a possible solution is to guarantee that the message *m2* or *m3* does not arrive before *m1* is finished processing.

Chapter

7

Design and Modeling Framework

This chapter discusses the major implementation issues one may encounter while developing an environment to support the activities presented in the book. The discussion centers around the construction of a design and synthesis environment (DASE), which is the supporting framework for DSL. The environment grew from a collaboration between McGill University's Micro-electronic and Computer Systems (MACS) laboratories and Bell Canada's Quality Engineering and Research group. It accepts a DSL model as input and provides capabilities for model configuration, library support, module verification, design exploration via simulation, and module synthesis to a subset of VHDL. The development of the environment was motivated by a desire to support a top-down structured approach to design.

The major components of DASE are shown in Fig. 7.1. Specifications are entered through a user interface and captured by the environment through the use of the internal representation, DSL. The language is interpreted through a DSL processor, which manages the library and DSL model configurations. The library support is a key component in permitting model reuse and facilitating rapid prototyping. The DSL processor also interacts with a DSL simulator and has provisions for a predicate/transition net–based analysis tool (such as PROD, Gronberg 1993) to provide verification of module properties. The DSL simulator interacts with the DSL processor during design exploration activities to refine the model as required. The final component is the synthesizer, which translates the (refined) hardware DSL models into a behavioral VHDL representation.

This chapter will present the various components of DASE and describe their functionality.

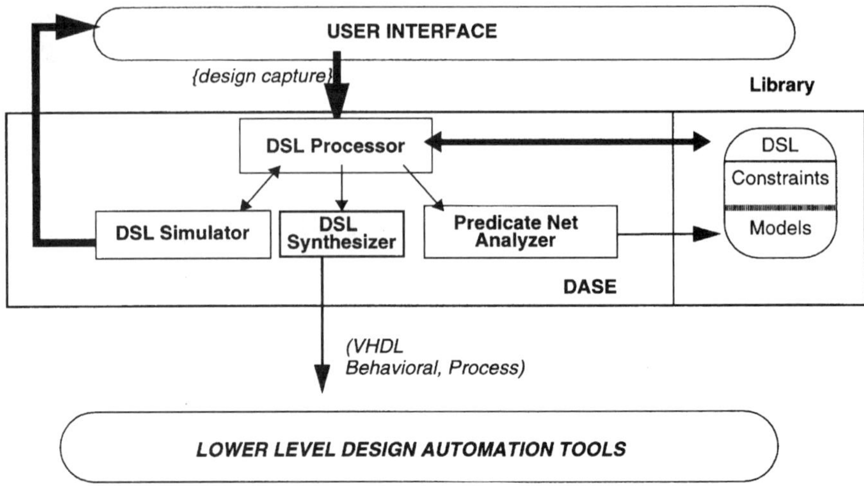

Figure 7.1 Major elements of DASE *(1995 IEEE)*.

Input Formats

A design framework will accept input at a certain level of abstraction. In the case of one supporting DSL, various forms of input may be possible. For example, the high-level description of parts of a design may be given in any one of the representational forms discussed in the earlier chapters. Ideally, the user should have facilities to model at a graphical level close to the conceptual design. In reality, this may not be easy to accomplish. The choice of the input representation can be based on diverse factors, such as the familiarity of designers with a language, the availability of tools, or corporate policy. Consequently, support for the translation of the different input formalisms to the intermediate language of the design environment will be required in many cases. The construction of the translators is analogous to that of the parser described in the synthesis section later in the chapter.

As a basis, DASE permits users to interact through a command-line interface. The construction of domain-oriented graphical interfaces is beyond the scope of this book. However, the command-line interface permits access to the DSL constructs as well as to the environment-specific elements described in this chapter.

DSL Processor

The DSL processor is a central component which manages the DSL models and is the primary interface with the user. Its main functions are the management of library modules, constraint checking (with the simulator), and module communication support (such as determination of unspecified destination modules). The library support system of DASE provides a high-level means for the modeling and management of models for telecommunication systems. The support entails the organized storage and retrieval of modules so that library modules are accessible to the environment as required. This implies that the DSL processor must be capable of accessing modules at different levels of representation and utilizing them as warranted. Design details which may be necessary, but are not of interest to a particular user, can be processed by the DSL processor transparently, freeing the designer to concentrate on the pertinent details. For example, if the modeler is modeling an application layer of a protocol, all lower-layer messages used within the model can be transparently maintained and reused through a library supporting the protocol. Alternatively, different levels of abstraction can also be included within the same model because of the ability of the environment to maintain a message hierarchy.

Library support

The top-down design approach of DSL is supported by an experimentation and model management mechanism through which reusable "generic" model components are made available to support simulation of different design options. A versatile library support system which enables creation, storage, and retrieval of module libraries in an organized manner is managed by the DSL processor. Libraries maintain all module information, as well as added rules regarding any constraints to be imposed on the modules, any configuration rules to be applied to the components of the library, and the interface specification of the library module. The interface specification is created by the library system to define exactly what ports are available for communication with the library module.

Constructing the hierarchy tree. Library modules are invoked with the *use_dsl_library([NAMES])* DSL statement defined in the previous chapters. Upon invocation, each library module is processed in an order identified through a model hierarchy tree. The tree is created by the DSL processor to establish a relationship between modules. The library modules are processed breadth first by the DSL processor. This ensures that any specific requirements (such as interconnections and instantiations) are identified by the immediate parent module. Hence library module configurations are affected by the environment into which they are invoked.

An example of a model hierarchy is shown in Fig. 7.2, where the top-level module, module a1, utilizes two other library modules, modules b1 and c1. A tree is created, with module a1 identified as the root

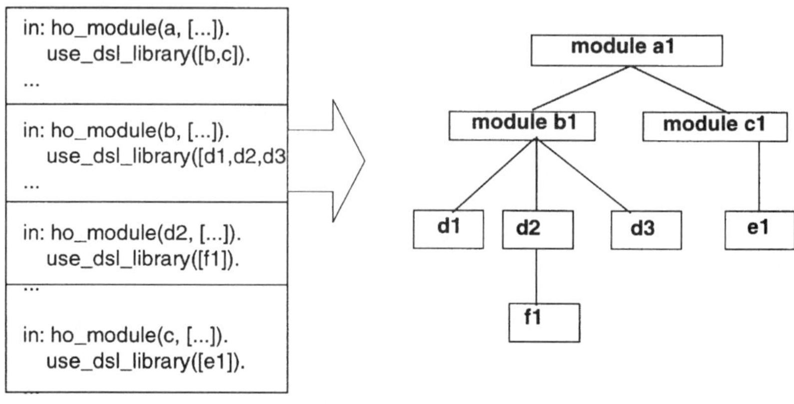

Figure 7.2 Model hierarchy example.

and modules b1 and c1 as subsequent nodes. The procedure is then repeated for the subsequent nodes to create the complete tree. Instantiations of module b1 are defined by module a1, which can affect the way in which b1 instantiates its subcomponents d1, d2, and d3.

The implementation of the hierarchy within DASE is accomplished through a set of five major Prolog predicates, which are shown in the code listing in Fig. 7.3. The *load_dsl* predicate initiates the hierarchy creation sequence by invoking the other four predicates described below. Initially only the main DSL model is loaded into the environment. Hence the root level of the tree is defined.

- The *set_up* predicate scans the modules at the root level of the hierarchy and initiates the data structures to maintain the tree. The *dsl_level(List, Level)* structure associates a level to a list of module names at the given level. The root level is identified as level zero.
- The *load_dsl1* predicate executes the *use_dsl_library* statements. Initially there will be one such statement (from the model at the root level). However, as the statement is executed, successive statements can be introduced into the environment from the library modules that are accessed. This reflects the breadth-first creation of the tree. The predicate *level_pred* identifies all new models created at the current level of the tree (identified by *dsl_level(Level)*) and records their levels within *model_level(Name, Level)*. Finally, the *process_lib(List)* predicate loads required new library modules, which will constitute elements of the next level in the tree. The predicate will eventually fail and force Prolog to backtrack and

```
load_dsl:- set_up, load_dsl1, level_make, scan_libs.

set_up:- ho_module(X, Y),
    extract_name(X, NAME),
    create_lib_list(NAME),
    create_loaded, fail.
set_up:- write('Top level modules identified'),nl.

    extract_name(X, N):- X =.. Listname, prune_list(Listname, N).
        prune_list([Head|Tail], Head).

    create_lib_list(N):- not(member(N,[name_nil])), retract(library_names(X)),
        append(X, [N], NN), asserta(library_names(NN)).

    create_loaded:-library_names(NN),asserta(loaded_lib_names(NN)),asserta(dsl_level(NN,0)).

load_dsl1:-
    use_dsl_library(List),
    level_pred,
    process_lib(List).
load_dsl1:- write('Finished loading libraries...'), nl.

    level_pred:-!,
        increment_level,
        make_model_levels.

        increment_level:-!,
            dsl_level(X),
            Y is X+1,
            retract(dsl_level(X)),
            asserta(dsl_level(Y)).

        make_model_levels:-
        make_modules,
        make_ho.

            make_modules:-
                module(X, _),
                not(model_level(X, XX)),
                dsl_level(L),
                asserta(model_level(X, L)),fail.
            make_modules:-
                write('updated module levels '),nl.

            make_ho:-
                ho_module(X,_),
                not(model_level(X, XX)),
                dsl_level(L),
                asserta(model_level(X, L)),fail.
            make_ho:-
                write('updated ho levels...'),nl.
```

Figure 7.3 DASE implementation of hierarchy.

reexecute *load_dsl1* for the next level. Once *load_dsl1* has completed, all library modules will have been loaded into the environment.

- The *level_make* predicate examines all the *use_dsl_library(Level)* statements to identify the levels for the library modules within *dsl_level(List, Level)*. It should be noted that the *use_dsl_library*

```

    process_lib(List):- get_first_element(List,First,Rest),
        prune_lib_list(First),
        process_lib(Rest).

        prune_lib_list(First):-
            loaded_lib_names(LIST),
            not(member(First, LIST)),
            load_library(First),
            write(' Loading library component '),
            write(First),
            write(' into current environment...'),nl,nl,
            retract(loaded_lib_names(L)),
            app(L, [First], NEW),
            asserta(loaded_lib_names(NEW)).

        prune_lib_list(First):-
            loaded_lib_names(LIST),
            member(First, LIST),
            remove_element(LIST, First, NEW),
            app(NEW, [First], Newlist),
            retract(loaded_lib_names(XXX)),
            asserta(loaded_lib_names(Newlist)).

            load_library(DSL_NAME):- write('consulting library module '),
                write(DSL_NAME),nl,
                consult(DSL_NAME).

level_make:-
    use_dsl_library(List),
    update_count(List,-1), fail.

level_make:- write('Built hierarchy tree...'),nl.

    update_count(List, Newcount):-
        dsl_level(X, Count),
        not(dsl_level(List,XXX)),
        Count > Newcount,
        Count2 is Count+1,
        update_count(List, Count2).

    update_count(List, N):-!,
        not(dsl_level(List, XXX)),
        N =\= -1,
        asserta(dsl_level(List, N)).

scan_libs:- use_dsl_library(X), add_to_lib_names(X), fail.
scan_libs:- nl.

    add_to_lib_names(New):- !, retract(library_names(List1)),
        app(List1, New, Result),
        asserta(library_names(Result)).
```

Figure 7.3 *(Continued)*

statements are maintained in the order in which they were invoked (root modules first).

- For convenience, *scan_libs* creates a structure *library_names(List)* which records the list of library modules loaded in the environment.

Library module construction. A means for generating library modules is needed. Library modules are created using the *build_dsl_library(NAME)* predicate interpreted by the DSL processor, which interactively constructs the necessary data structures for the library module *NAME*. The *NAME* is a valid module name loaded within the environment or available as a file name in a model directory. The basic structure of a library module is depicted graphically in Fig. 7.4. The predicate proceeds through three phases of construction of a library module structure.

- *Phase 1.* The first stage is to create the module and higher-order module definitions of the library. The system searches for the specified module and, if successful, it re-creates the behavior of the module (the predicate will fail in the case of an undefined module).
- *Phase 2.* The second stage is the construction of the library configuration rules and starting conditions, which are created interactively. The system provides a template which, when completed by the user, specifies the nature and use of the module. The DSL processor will examine the module behavior and identify (to the user) any resources and path names used by the module that

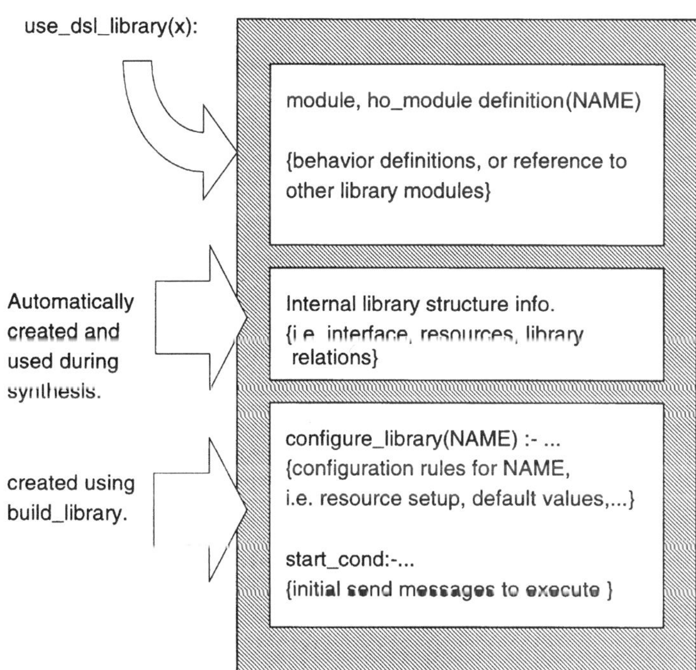

Figure 7.4 Library module structure.

require special treatment during configuration. The configuration rules are stored as Prolog rules, with the predicate head defined as

```
configure_library(NAME).
```

Configuration rules allow for the creation of module resources, interconnection definition (for higher-order modules), definition of constraints, and instantiation of other modules. During library invocation the rules are executed by the DSL processor to create the required resources for the library module. The rules may be parameterized, allowing for different possible types of path definitions. The user may also define default values for parameters used by the library module, which is maintained as a default *(Module (D1,..., Dn)* predicate, where the parameters *D1* to *Dn* are default values the system is to use. Defaults are used only if the argument value is not defined by the environment during model configuration. Conflicting or missing interconnections during configuration are identified by the DSL processor and reported back to the user.

The structure of a configuration rule is shown in Fig. 7.5. The regions outlined by bold text are inserted by the DSL processor as required. Four lines (with italic text) are shown that require user input. These are definitions of resources (local variables, data structures), paths (interconnections), other modules, and user-defined subrules. The figure indicates only one line per definition; however, there can be multiple lines for each definition. It should

```
configure_library(NAME):-
        isa(INSTANCE, NAME),
        asserta(resource(INSTANCE, RESOURCE, VALUE)),
        asserta(path(INSTANCE, OTHER, [PORT, OTHER_PORT])),
        asserta(module(NAME2, [])),
        other_rules(...),
        fail.
configure_library(NAME):- write('Library module '),
        write(NAME), write(' configured\nl').
```

List of Variables:
NAME: name of the library module,
INSTANCE: an instance name for NAME extracted from the environment,
RESOURCE: name of resources for INSTANCE,
VALUE: initial value of RESOURCE,
OTHER: name of module(s) to interconnect with INSTANCE,
PORT and OTHER_PORT: port names for INSTANCE and OTHER,
NAME2: any other module instantiations that are dependent upon NAME.
other_rules(...): are further user defined sub-rules that are defined. The functor and arity of this predicate is not significant and is used merely here as an example.

Figure 7.5 Configuration rule structure.

also be noted that the configuration rules are applied to all instances of *NAME*.

Start conditions are also identified in this phase of library construction. A rule beginning with the *start_cond* predicate identifies any messages that need to be sent to the module to place it in a valid state of operation. This allows some control over the initial starting state of modules. All *start_cond* rules are executed in turn during library invocation, scheduling any initial messages for the simulator.

- *Phase 3.* The third stage is to capture the interface and resource information for the module. This is achieved by the DSL processor by grouping the required module resources, required ports, and relations. In the latter case, a class relationship (if applicable) is associated with the module library and is established with the *isa* predicate as with modules. The class relationships for a model library are maintained within a file called *lib_def*. The file consists of *isa* predicates identifying any relationships that may exist between library modules.

Figure 7.6 illustrates an example for building a library module (called x) that is of the same class as an existing module (y). The user, shown on the left-hand side of the figure, supplies the module's name and configuration rules and identifies a relationship to another module (y). The DSL processor will verify the existence of the library module and verify that all resources and paths used within module x's behavior have been properly defined.

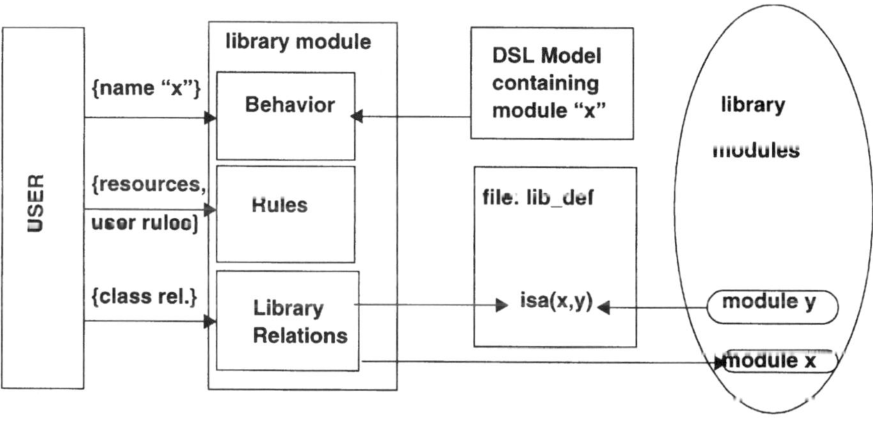

Figure 7.6 Example of constructing a library module.

The class relation allows library components to be identified and utilized by the system during simulation, design exploration, and synthesis.

Upon creation of a library module, DSL also verifies that constraints are not in conflict and that there is no inconsistency with the library interface and its associated modules. Library configuration rules allow a module to arrange itself depending upon different conditions during invocation. These rules allow a generic library module to be reused in different ways as the design requirements demand.

Implementation of library configuration

After the required library modules have been introduced into the environment, they are ready for configuration. This is achieved through a predicate called *lib_config* as follows:

```
lib_config:-library_names(LIST), lib_config(LIST).
```

lib_config(LIST) will process all the library names and configure them according to the configuration rules. The Prolog code for this is

```
lib_config([NAME|REST]):-
    configure_library(NAME),    /* execute the library configuration
                                rules for NAME */
    dsl_level(LIST, LEVEL),     /* check the level in the tree for
                                NAME */
    member(NAME, LIST),
    Lev2 is LEVEL + 1,          /* increment the level number */
    retract(dsl_level(XXXX)),
    asserta(dsl_level(Lev2)),   /* change the current level to the
                                new one */
    make_model_levels,          /* create model levels for any new
                                modules */
    lib_config(REST),           /* process the remaining names */
lib_config([]):-write('…All Library modules configured…'),nl,nl.
```

The basic component of the predicate is the execution of the *configure_library* rules. A side effect of this may be the creation of new modules (instances of library modules). Hence there is an additional need to invoke the *make_model_levels* (which was shown in the previous figure) predicate to identify the levels of the new entities in the hierarchy.

Implementation of DSL statements

Within the DASE framework, the interpretation of DSL statements is performed within the DSL processor. This component could have also been implemented as part of the simulator. The rationale within DASE was to separate the scheduling components from the language directives into the respective functional constructs.

Send statement. The *send* statement can be utilized in four different cases, as described in Chap. 4. Consequently, all the different possibilities must be interpreted. The implementations are not very different in nature, and may require an additional predicate. The implementation described here will cover most of the communication possibilities. A complete code description of the *send* statements can be found in the DSL code listing in the appendices.

The goal of a *send* statement is for a given message to be executed by a valid recipient module. For now, we will assume that a predicate exists to decode a module's behavior and identify a *send* statement for execution. The Prolog description of the *send* statement is as follows:

```
send(Destin,S_port,Message):-
current_module_processed(Source),
int_mod_type(Source, Destin, S_port, Sender, Rec, SPort),
mod_type(Sender, Rec, SPort, Module),
isa_hierarchy(Module, Module 1),
module(Module 1, MethodList),
get_method(Message, MethodList, Method),
queue message(Module, Message).
```

To fulfill its goal, the code first identifies the module which initiated the *send* statement. This is assumed to be maintained within the *current_module_processed(Source)* structure. The *int_mod_type* (interpret module type) predicate determines whether the destination is a module or a higher-order module. If a higher-order module is uncovered, a final destination is resolved that is a module. This is necessitated by the fact that higher-order modules provide structural information; hence any behavior to be executed at the destination can be derived only from the behavior list of a module.

When a valid module is detected, its behavior is obtained as a list named *MethodList*. The predicate *get_method* is used to extract the behavior associated with the message from *MethodList*. This is then placed on the appropriate module's input queue. Figure 7.7 depicts the Prolog description of the supporting predicates.

Delay. Time attributes associated with the execution of modules are represented by the DSL *delay* statement. When a *delay* statement is encountered, the module involved is in one of two possible states: *ready* or *delayed*. In the former case, the module is ready to execute statements that are associated with a message containing the *delay* statement. However, on encountering the *delay* statement, the module will be placed in a *delayed* state for a period of time determined by the *delay* statement's parameter. The message will therefore be scheduled for a later period of time and kept on top of the module's input queue.

```
                int_mod_type(Source,Destin, Port, Source, Destin, Port):-
                                path(Source,Destin,[Port,_]),
                                module(Destin,_).

                int_mod_type(Source,Destin, S_port, Parent, Target, TP):-
                                path(Source, Destin, [Port, D_port]),
                                ho_module(Destin, List),
                                member(Source, List),
                                path(Source, Destin, [Port, D_port]),
                                path(Destin, Next, [D_port, N_port]),
                                not(equalt(Source,Next)),
                                not(member(Next, List)),
                                int_mod_type(Destin, Next, D_port, Parent, Target, DPort).

                int_mod_type(Source,Destin, S_port, Destin, Next, D_port):-
                                path(Source, Destin, [Port, D_port]),
                                ho_module(Destin,_),
                                path(Destin, Next, [D_port, N_port]),
                                int_mod_type(Destin, Next, D_port, Dest, New, Port_D).


mod_type(Source,Dest, Port, Dest):- module(Dest, _).

mod_type(Source,Dest, Port, Module):- ho_module(Dest, _),
     path(Source, Dest,[Port,Dest_port]),
     path(Dest,Module,[Dest_port,_]),
                                 mod_type(Dest, Module, Dest_port, Module).


isa_hierarchy(Module,Module).

isa_hierarchy(Module,Module1) :-
  isa(Module, Module2),
  isa_hierarchy(Module2, Module1).


get_method(Message,[FirstMethod|Rest], Method) :-
  decode(Message, FirstMethod, Method).

get_method(Message, [_|Rest], Method) :-
  get_method(Message, Rest, Method).


     decode(Message, Message, true).

          decode(Message,(Message:-Body), Body).
```

Figure 7.7 Prolog specification of *send* statement.

When the message is next processed, it will be in a *delayed* state. In this case, the *delay* statement is ignored, the state of the module is reset to *ready,* and subsequent DSL statements following the *delay* statement are executed.

In the implementation, the flow control is managed by the use of *module_state(STATE)* predicates. When *STATE* is set to *ready,* the module is in a normal mode of operation (i.e., its operation has not been suspended). If *STATE* is *delayed,* the module is suspended and will awaken when the delayed time has elapsed. The Prolog representation of the *delayed* case is simply

```
delay(Delta):- current_module_processed(Module),
   module_state(Module, delayed), !,
   retract(module_state(Module,_)),
   asserta(module_state(Module, ready)).
```

However, for the case when the statement is first encountered and the module's operations are suspended, some additional logic is required, as below:

```
delay(Delta):- current_module_processed(Module),
   sim_time(T1), retract(module_state(Module,_)),
   asserta(module_state(Module,delayed)),
   retract(next_time(Module,_)),
   asserta(next_time(Module,T1 + Delta)),!, fail.
```

The *next_time* (refer to Chap. 5 for data structures for the simulator) of the module is updated to reflect the scheduled time at which the module will resume operation. A *fail* predicate at the end ensures that any statement following the delay will not be executed until the future time period.

Resource manipulation. The primary resource manipulation statements, *set_res* and *check_res,* are equivalent to assert and query operations in Prolog. The *set_res* statement can be described in Prolog as

```
set_res(Resource_Name,Value):-current_module_processed(Module),
      retract(resource(Module, Resource_Name, Old_value)),
      asserta(resource(Module, Resource_Name, Value))
```

In the description, the old value of the resource is discarded and replaced by a new parameter. The *check_res* statement is simply a query, such as

```
check_res(Resource, Value):-current_module_processed(Module),
     resource(Module, Resource, Value).
```

The value of the resource is then bound to the variable within the *check_res* statement.

The *remove* and *create* statements can also be described in a similar manner as

```
remove(R, V):-current_module_processed(Module),
    retract(resource(Module, R, V)).
```

and

```
create(R, V):- current_module_processed(Module),
    asserta(resource(Module, R, V)).
```

A final statement worth mentioning here is *probe*. This statement returns data associated with a persistent port. This is a special case of a resource query, where the resource is determined from a module's name. The name is derived from the port in connections. The Prolog description below reflects these concerns, with the query using a *path* and then a *resource* predicate:

```
probe(PORT, DATA):-current_module_processed(Module),
   path(Module, ModuleX, [PORT, PortX]),
   resource(ModuleX, persistent_port, (PortX, DATA)).
```

DSL Simulator

The DSL simulator provides a means for experimentation and simulation of the DSL models. The simulator interacts with the DSL processor and library, and manages the scheduling and exchange of messages between modules and the update of a virtual simulation clock, as well as ensuring that no constraints are violated.

The simulator utilizes several data structures to schedule messages (shown in Fig. 7.8). Messages are scheduled and maintained in *message_queues* created by the environment for each module. The queue is a list of tuples consisting of the message to be executed at the module and its creation time (to be used for statistics and constraint checks).

Creation of *message_queues* and other data structures can be achieved through a predicate such as *initialize:*

```
initialize:- module(Name,_), not(isa(_,Name)),
          asserta(message_queue(Name, [])),
          asserta(next_time(Name,-1)),
```

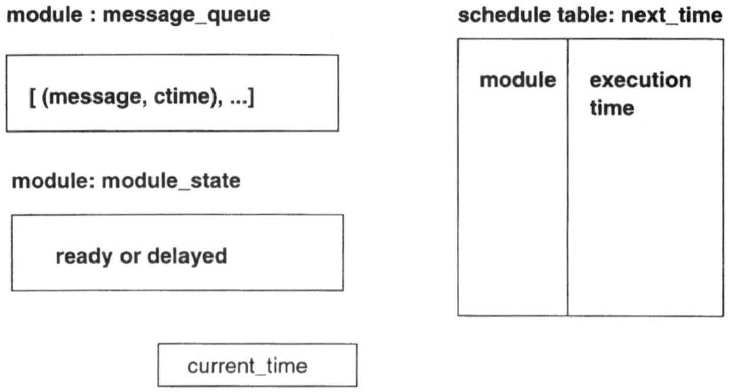

Figure 7.8 Simulator data structures.

```
asserta(module_state(Name,ready)),
asserta(path(Name,Name,[in,out])),
fail.
```

The predicate traverses each instance of modules and instantiates the supporting data structures. The simulator also defines modules to be in a *ready* or *delayed* state, which is maintained in a variable *module_state*. If the module is in the *ready* state, the next message in the module's queue can be processed, whereas the *delayed* state indicates that the module is delayed until a specified time period. This time period is maintained in a schedule table named *next_time*. The simulator schedules the messages based upon this time value. The *initialize* predicate will set the *next_time* of each module to -1 by default. This indicates that there are no messages scheduled for the module.

When a *send* statement is executed as part of a module's behavior, the simulator automatically checks the *next_time* table to determine if any messages are scheduled for the destination module. This is implemented as a Prolog predicate *queue_message*. If no messages are scheduled (a negative time entry in the *next_time* table), the execution time for the module is updated to the *current_time*. In either case, the message and its creation time (*current_time*) are appended to the *message_queue* for the destination module. The corresponding Prolog code for these operations is as follows. The predicate *queue_message* can be described by

```
queue_message(Mod1, Behavior):-
                sim_time(Time),
                retract(message_queue(Mod1,Elements)),
                app(Elements, [(Behavior,Time)], New_q),
                asserta(message_queue(Mod1, New_q)),
                update_next(Mod1).
```

The predicate *update_next* identifies whether the new message is the only message in the queue. If it is, the *next_time* is updated to the current simulator time (*sim_time*). This is given by

```
update_next(Modulex):-
         next_time(Modulex, Temp),
         Temp = = -1,
         sim_time(Time),
         retract(next_time(Modulex, _)),
         asserta(next_time(Modulex, Time)),print('\n').
      update_next(Modulex).
```

The simulator will then process all messages with execution time equal to the *current_time*. The processing involves removing the message from the respective *message_queue* and executing it through the DSL processor. If the message was in a *delayed* state, then the *delay* statement within the behavior is bypassed, allowing for the continuation of execution of the module's behavior, and the *module_state* is set to *ready*. If the module was in a *ready* state and a *delay* statement is encountered, then the simulator places the module in a *delayed* state and updates the respective time in the *next_time* entry.

The implementation of the above steps is given in Fig. 7.9. The main predicate in the figure is the *queue_handler*, which performs the above queue management operations (removing the messages to be processed in the current simulation cycle). The *get_method* predicate will extract the message to be processed from the appropriate module. Another predicate (*test_and_exec*) executes the DSL statements that make up the message. This predicate is expected to fail if a *delay* statement is encountered, so that statements following the *delay* statement are not immediately processed. The second *test_and_exec* predicate handles the case of the predicate failing. It simply acknowledges that the module is in a *delayed* state.

During each behavior execution, any constraints associated with the module are checked. This is indicated by invocation of the *check_constraint* predicate. If a constraint is violated, the simulator

```
queue_handler:- sim_time(Main_time),
                next_time(Modx, Main_time),
                retract(current_module_processed(XMODULE)),
                asserta(current_module_processed(Modx)),
                message_queue(Modx,[(Message,Qtime)|Rest]),
                isa_hierarchy(Modx, Module1),
                isa_hierarchy(Message, Mess),
                module(Module1, MethodList),
                get_method(Mess, MethodList, Method),
                next_time(Modx, Main_time),
                test_and_exec(Method,Mess,Modx).

test_and_exec(Method,Mess,Modx):-
                call(Method),!,
                check_constraint(Modx, Mess),
                remove_first(Modx).

test_and_exec(Method,Mess,Modx):-
                module_state(Modx, delayed).

/* The following predicates remove the top message from a module's message queue. A check is also
made
to determine if the queue is empty. If so, a -1 value is inserted in the next_time predicate for the module
*/

remove_first(M):- module_state(M,ready),
                retract(message_queue(M, [First|Rest])),
                asserta(message_queue(M, Rest)),
                null_check_res(M, Rest).


null_check_res(Module,[]):- retract(next_time(Module,_)),
                asserta(next_time(Module,-1)).
null_check_res(Module,X).
```

Figure 7.9 Implementation of scheduler.

will identify the violation to the user. Included in the message sent to the user is a list of possible alternative library modules of the same class. This is retrieved through the *lib_def* file maintained by the DSL processor. The designer can then choose to alter the design with the replacement, relax the constraint, or ignore the constraint.

The simulator determines the next highest time possible in the *next_time* table and sets the simulator clock (*current_time*) to it. The previous step is then repeated. An implementation of these operations is given by the *update_sim_clock* predicate listed in Fig. 7.10.

The model hierarchy created during library invocation is also utilized by the simulator to aid the designer through design exploration. An observation level, which represents the level of detail different modules define, is defined by the simulator. For example, the root of the model hierarchy is labeled as level 0, which represents the most abstract module. The next level down would consist of modules that are referenced and inherited by the topmost module, and similarly for subsequent levels.

```
manipulate_clock:-
                sim_time(Main),
                next_time(Modx, Main).
manipulate_clock2:-
                update_sim_clock.

update_sim_clock:- next_time(_, Time1),
                max_sim_time(MaxTime),
                sim_time(Main),
                MaxTime >= Main,
                Time1 >= 0,
                find_low_time(Time1, Next_time).

update_sim_clock:-
                sim_time(Main_time), system('clear'),nl,
                write('—> CURRENT SIMULATOR TIME is '),
                write(Main_time),write(' microseconds.'),nl, XT is cputime,
                write('=====================================================
'),write(XT),
                nl,nl.

test_sim_time:-
                max_sim_time(MaxTime),
                sim_time(Main),
                (Main >= MaxTime,
                retract(terminate(X)), asserta(terminate(true)), end_sim; true).


find_low_time(Curr_time, Temp):- next_time(_, Temp),
                Temp >= 0,
                Curr_time > Temp,!,
                find_low_time(Temp, Xt).

find_low_time(Curr_time, Curr_time):- retract(sim_time(Y)),
                asserta(sim_time(Curr_time)).
```

Figure 7.10 Implementation of clock update mechanism.

Abstracting hierarchy has often been employed to facilitate simulation and modeling at various levels of detail (Zeigler 1984). For example, a system-level analyst may not be interested in the lower-level details of a given model, whereas a hardware designer would be more concerned with the lower timing details of the simulation. The DSL simulator provides for dynamic alteration of the level of abstraction viewed by defining an observation level. The level can be defined to be limited to a given level or to be a particular level and all levels below it.

The observation level can be set at any time in the simulator with the *set_level(LEVEL)* command. An implementation of this is

```
set_level(X):-retract(dsl_level(Y)),
asserta(dsl_level(X)),
write('DSL SIMULATOR: Observable level set to'),write(X),nl.
```

The observation level is assumed to be retained within a Prolog fact *dsl_level(X),* where *X* is the level the simulator is set to. By default, the simulator is initiated to display all levels of information. The simulator also utilizes a *list* command to provide snapshots of the state of the system. This command is in a menu-driven form which permits viewing of resources, modules and higher-order modules, states of the simulator data structures, and the interconnections of modules. All of the views can be constrained to a particular level or range of levels of observation. The typical output of the command is shown in Fig. 7.11. The level of the observed module or data structure is displayed in the first column. The sample shows four particular views of the state of the simulation. The first is a display of the hierarchy. As an example, a module

```
Hierarchy listing
delivery_q(1)        --> delivery_queue(1)
priority_q(1)        --> priority_queue(1)
aclock(3)            --> generic_clock(3,2,44)

Listing state of the module queues:
LEVEL  Module:           Messages queued:          Delayed until time:
2      clock(1,3)   [(clock_count(44) , 0.0)]     0.0
2      clock(1,2)   [(clock_count(44) , 0.0)]     0.0

Listing Resources Created:
LEVEL  Module:           Resource:                 Variable(s):
4   priority_q(3) p_queue(1)                []
4      gbcast(2)         label_count              1

Listing Path interconnections:
2      output(4,1) is connected to:
       --> output(4,1)      Port pair:  [in,out]
       --> switch(4)  Port pair:  [out_port(4,1),outline(4,1)]
       --> buffer(4,1)      Port pair:  [queue_port(4,1),out_port(4,1)]
       --> atm(4,_764)       Port pair:  [out_port(4,1),outline(4,1)]
```

Figure 7.11 Typical output of *list* command.

delivery_q(1) is defined to inherit behavior from another module, *delivery_queue(1)*. The second and third views show the state of the message queues and resources for each module. A module *clock(1,3)* is indicated as ready (a delay time of *0.0* is shown in the rightmost column) to process a new message, *clock_count(44),* from its message queue. The final view is a description of the interconnections between the modules.

The implementation of the *list* command follows a typical Prolog input/output predicate structure. A complete listing of the command implementation is given in Fig. 7.12. Further elements of interest for viewing similar to the existing ones can be added to the menu.

After DASE loads the library modules into the model, an autogeneration of higher-order modules is necessary. This is due to the probability of some higher-order modules being defined simply as *ho_module(Name, [])*. This type of notation is common in cases where a referenced higher-order module is an instance of a library module. In such cases, the submodules contained within the higher-order module must be determined and specified within the environment, and the associated paths between such elements must be created. In Prolog terms, a predicate (*ho_module_lib_config*) can be used to initiate the two actions as below:

```
ho_module_lib_config:-
      ho_module(X, []),
      isa(X, OBJECT),
      ho_module(OBJECT, SUBS),
      process_ho_list(X, SUBS),
      ho_path_config(X, OBJECT),
      ho_module_lib_config.
ho_module_lib_config.
```

For each higher-order module with an empty list of submodules, instances of the higher-order module are found. The predicate *process_ho_list* is used to create the submodule instances, and *ho_path_config* is used to define the paths. The Prolog description of the former is given as

```
process_ho_list(X, PARAMETERS):-
       retract(ho_module(X, [])),
       asserta(ho_module(X, PARAMETERS)).
```

The second predicate establishes the paths defined between the higher-order module and its submodules. Two cases for the predicate are needed in order to account for bidirectional connections. The third case of the predicate simply returns control to the main predicate.

```
ho_path_config(X, OBJECT):-
      path(OBJECT, D1, Z1),
```

```
list:- system('clear'),write('DSL SIMULATOR: Interactive List mode \n'),
        write('Do you wish to limit the scope of list to only one level? (y/n) '),
        get(S),error_check(S, SS), char_code(MODE,SS),system('clear'),
        write('\nPlease type an appropriate integer to indicate level of observation: \n'),
        get(L),char_code(LEV, L),make_number(LEV,LEVEL),
        write('Observation level for list is set to '),write(LEVEL),nl,nl,
        write('What do you wish to list?\n'),
        write('r - resources\n'),
        write('h - ho-modules\n'),
        write('i - hierarchy\n'),
        write('s - state of queues\n'),
        write('p - path connections\n'),
        write('q - quit'),
        write('? '),
        get(OP), char_code(OPTION,OP),system('clear'),
        nl, list(MODE, OPTION, LEVEL).

list(X, q,L).
list(XX,i,L):-write('Hierarchy listing \n'),
        isa(X, Y),
        write(X), write(' \t\t -> '), write(Y),nl,fail.

list(y,r,L):-write('Listing Resources Created: \n'),
        write('LEVEL\tModule: \t\tResource: \t\t\tVariable(s): \t'),nl,nl,
        resource(NAME, RESNAME, PAR), otest_level(L,NAME),
        write(NAME),write('\t\t'),write(RESNAME),
        write('\t\t\t'), write(PAR), nl,fail.

list(y,s, L):- write('Listing state of the module queues:\n'),
        write('LEVEL\tModule: \t\tMessages queued: \tDelayed until time:\n'),nl,
        message_queue(M, Q), otest_level(L,M),
        next_time(M, T),
        write(M),
        write(' \t'),
        write(Q),
        write('\t'),
        write(T),nl,
        fail.

list(y, h, L):- write('List of HO_Modules:'),nl,write('LEVEL\tHO-MODULE NAME: \t\t [ Sub modules ]:\n'),nl,
        ho_module(X,Y), otest_level(L,X),
        write(X), write('\t'), write(Y), nl,fail.

list(y, p, L):-write('Listing Path interconnections:\n'),
        model_level(A,_),
        otest_level(L, A),
        write(A), write(' is connected to:\n'),
        list_the_paths(A).
```

Figure 7.12 Implementation of *list* command.

```
        asserta(path(X, D1, Z1)),fail.
ho_path_config(X, OBJECT):-
        path(D1, OBJECT, Z1),
        asserta(path(D1, X, Z1)),fail.
ho_path_config(X, OBJECT).
```

Additional environment commands are needed to activate and control the simulator time advancement. These can be implemented in

```
list(n, p, L):-write('Listing Path interconnections:\n'),
                model_level(A,_),
                test_level(L, A),
                write(A), write(' is connected to:\n'),
                list_the_paths(A).

list_the_paths(A):-
                path(A,B,Y),
                write('\t --> '),write(B),write('\t Port pair:  '),write(Y),nl,fail.

list(n, r,L):- write('Listing Resources Created: \n'),write('LEVEL\tModule: \t\tResource: \t\t\tVariable(s):
\t'),nl,nl,
                resource(NAME, RESNAME, PAR), test_level(L,NAME),
                write(NAME),write('\t\t'),write(RESNAME),
                write('\t\t\t'), write(PAR), nl,fail.

list(n, s, L):- write('Listing state of the module queues:\n'),
                write('LEVEL\tModule: \t\tMessages queued: \tDelayed until time:\n'),nl,
                message_queue(M, Q), test_level(L,M),
                next_time(M, T),
                write(M),
                write(' \t'),
                write(Q),
                write('\t'),
                write(T),nl,
                fail.

list(n, h, L):- write('List of HO_Modules:'),nl,write('LEVEL\tHO-MODULE NAME: \t\t [ Sub modules ]:\n'),nl,
                ho_module(X,Y), test_level(L,X),
                write(X), write('\t'), write(Y), nl,fail.

list(res):- list(r,10).
list(s):-list(s, 10).
list(h):-list(h, 10).

list:-nl,write('done.'),nl.
list(X,Y,Z):-nl,write('done.'),nl.

test_level(L,NAME):-
                model_level(NAME, LEVEL),
                L>=LEVEL,write(LEVEL), write('\t').
test_level(L,NAME):-
                model_level(NAME, LEVEL),
                L=LEVEL,write(LEVEL), write('\t').
```

Figure 7.12 *(Continued)*

many different ways. Here the implementation consists of using a *step* and *start* predicate. The former permits manual definition control over how many time steps the simulator clock can advance. The latter permits a free-running simulator, stopping only when the maximum simulator time (defined by the user) is reached or an unexpected error condition occurs. A *reset* predicate is also useful for reinitializing the simulator to prepare for another simulation run.

The *step* predicate is appealing for situations in which a given time period is of interest or for debugging various event interactions. A possible implementation for this is

```
step:- sim_time(Main_time),
       write('DSL SIMULATOR: <<update>> CURRENT SIMULATOR TIME
       is'),
       write(Main_time),
       queue_handler.
```

The *queue_handler* processes all messages ready for execution in the current time period. The description of the *step* command is simple and of limited use, since there is no simulator clock update involved. A more useful implementation would provide such facilities. For example, the following listing is a possible way of providing additional control over the simulator clock:

```
step(0).
step(X):- step, XX is X-1, update_sim_clock, step(XX).
```

The code uses a parameter as part of a *step* command to increment the clock by the specified number of clock ticks. For example, *step(5)* would update the simulator clock five times, each time executing the messages that are to be processed. The user can advance the simulation at prescribed time periods.

The *start* command can be viewed as a special case of the *step*. It is simply a sequence of *step* commands that progresses until the end of simulation time is reached (this activates a predicate *end_sim* to inform the user of the state). A listing of the code for the predicate is given in Fig. 7.13.

Synthesis

The synthesis process is invoked through the *synthesis* command within DASE. The environment utilizes various data structures to effectively synthesize a design to VHDL. The relevant support information required by the environment in order to proceed through a successful synthesis is shown in Fig. 7.14.

The environment maintains a list of module names that are to be synthesized. The list is extracted from the *module_type(NAME, hardware)* predicates (see Chap. 6). The predicate identifies that the module will be implemented as a hardware component. DASE will then extract all pertinent predicates regarding the module(s) into an intermediate file (*file_dsl*) as guidance to the parser. The predicates are created by the *synthesis* command and can consist of information such as

1. The complete module behavior, which is obtained directly from the module definition and the *isa* statements.
2. The *outport_list(Module, OUTPort, [Behavior]),* which identifies all messages that *Module*'s output port *OUTPort* will support. The

```
start:- sim_time(Main_time), system('clear'),nl,
                    write('—> CURRENT SIMULATOR TIME is '),
                    write(Main_time),write(' microseconds.'),nl, XT is cputime,
                    write('===========================================================
'),write(XT),
                    nl,nl,
                    repeat,
                    queue_handler,
                    not(manipulate_clock), manipulate_clock2,test_sim_time.

start:- end_sim.

start(X):- retract(max_sim_time(_)), asserta(max_sim_time(X)), start.
go:-XCPU is cputime, asserta(begin_time(XCPU)),repeat,start,terminate(true), reset_streams.

end_sim:- nl,
                    write('======= Message from DSL simulator ==============='),
                    nl,
                    write(' No further messages to process: '),nl,
                    write(' Current simulated time is '),
                    sim_time(T),
                    write(T),write(' microsecs'),nl.


queue_handler:- sim_time(Main_time),
                    next_time(Modx, Main_time),
                    retract(current_module_processed(XMODULE)),
                    asserta(current_module_processed(Modx)),
                    message_queue(Modx,[(Message,Qtime)|Rest]),
                    isa_hierarchy(Modx, Module1),
                    isa_hierarchy(Message, Mess),
                    module(Module1, MethodList),
                    get_method(Mess, MethodList, Method),
                    next_time(Modx, Main_time),
                    test_and_exec(Method,Mess,Modx).

test_and_exec(Method,Mess,Modx):-
                    call(Method),!,
                    print_sim_time,
                    print('DSL PROCESSOR: Message '),print(Mess),
                    print(' completed at module '),printl(Modx),
                    check_constraint(Modx, Mess),
                    remove_first(Modx).

test_and_exec(Method,Mess,Modx):-
                    module_state(Modx, delayed).
```

Figure 7.13 Implementation of *start* command.

pertinent information is extracted from the *send* statements within the module and added to the *behavior* list.

3. The *inport_list(Module, INPort, [Behavior]),* which is used to identify all the messages that can be received (*Behavior*) through *Module*'s input port *INPort.* This information is obtained through a comparison of the possible sources of messages to a module (found by an examination of the *path* statement) and the corresponding

```
debug_dsl:- list(n,s,10).

manipulate_clock:-
                    sim_time(Main),
                    next_time(Modx, Main).
manipulate_clock2:-
                    update_sim_clock.

update_sim_clock:- next_time(_, Time1),
                    max_sim_time(MaxTime),
                    sim_time(Main),
                    MaxTime >= Main,
                    Time1 >= 0,
                    find_low_time(Time1, Next_time).

update_sim_clock:-
                    sim_time(Main_time), system('clear'),nl,
                    write('—> CURRENT SIMULATOR TIME is '),
                    write(Main_time),write(' microseconds.'),nl, XT is cputime,
                    write('==============================================
'),write(XT),
                    nl,nl.

test_sim_time:-
                    max_sim_time(MaxTime),
                    sim_time(Main),
                    (Main >= MaxTime,
                    retract(terminate(X)), asserta(terminate(true)), end_sim; true).
```

Figure 7.13 *(Continued)*

messages that can potentially be delivered to the respective port (this information is already available from the *outport_list*).

4. All *path* statements related to a module.

A precheck predicate verifies that all the requirements for synthesis to proceed are satisfied. One of the important checks is to identify and resolve any outstanding path interconnections. This is also the satisfaction of the third requirement for enabling synthesis (given in Chap. 6). The chosen solution within DASE is to identify a unique port name for each destination sharing the same output port of the source module. This is shown in Fig. 7.15. The diagram on the left depicts three modules connected to the same output port of module *a*. Such a connection is interpreted by the DASE synthesizer as the diagram shown on the right in Fig. 7.15. The respective destination names are appended to the output port name and instantiated as three separate ports. This operation is performed during synthesis whenever a multiple-destination *send* statement is encountered.

DSL-to-VHDL Parser

The *file.dsl* is input to a DSL parser. The parser is built using the UNIX *flex* and *yacc* utilities. The complete code listing of the parser

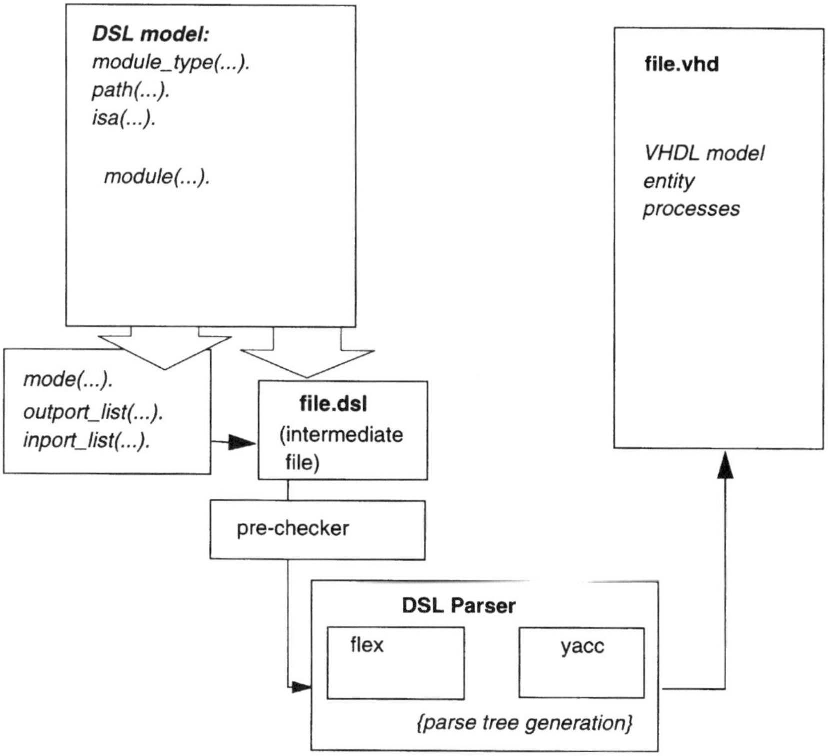

DSL model:
module_type(...).
path(...).
isa(...).
module(...).

mode(...).
outport_list(...).
inport_list(...).

file.dsl
(intermediate file)

pre-checker

DSL Parser
flex
yacc
{parse tree generation}

file.vhd
VHDL model
entity
processes

Figure 7.14 Synthesis environment in DASE.

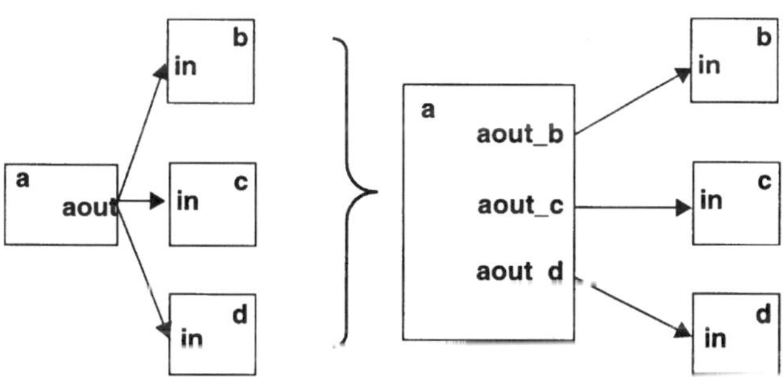

Figure 7.15 Port resolution in synthesis.

can be found in the appendices. As the input file is parsed, a parse tree (represented in a linked list structure) is created within the parser. The data structure is shown in Fig. 7.16. Each branch of the tree identifies the VHDL equivalent of DSL actions, variables, branching conditions, and other model information. Not all fields are filled dur-

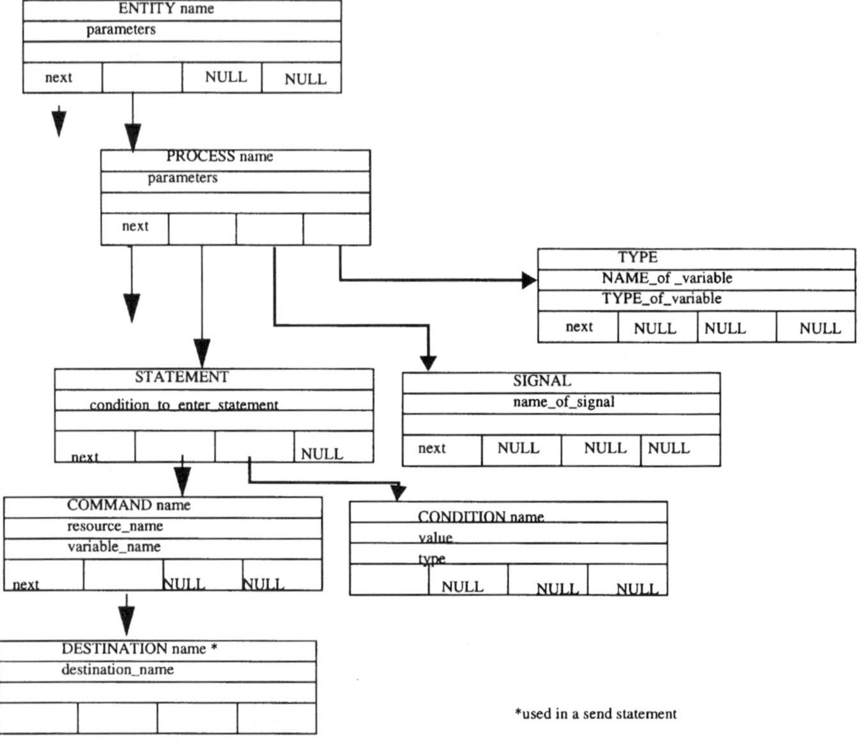

Figure 7.16 Parser data structure.

ing the initial parse of the input. A second pass through the tree entries is required to establish the VHDL counterparts to the DSL code. The parse tree is then traversed to generate the corresponding VHDL model.

The nodes of the parse tree contain three information fields. The first identifies the type of node. This can be an entity (module name), a command (such as *set_res* or *create*), a signal (identified VHDL signal names), a type (identified types), a statement (any branching information), or a condition node (any conditions before attempting commands). The other two information fields are used to store the relevant data with the corresponding node. Each node also has four pointers. One of these is dedicated to point to the next node; however, the others are used as warranted to indicate other nodes that hold information regarding the particular DSL statement's parameters.

During the course of the construction of the parse tree, the data within nodes are modified as the complete DSL code is parsed. The tree is then traversed, and translation rules (explained below) are applied to complete the tree and produce VHDL code.

DTSS example revisited

A design model is as good as the experimentation it supports. Different operational scenarios or configurations help in refining and studying the overall behavior of the modeled system. Design experimentation requires the definition of two models. The first is a model of the system under consideration. Such a model is studied, and different sections are eventually synthesized to hardware. The DSL modules presented all fall into this category. The second type of models that are required are those that emulate some key aspects of the environment that the system under consideration will work in. Such models can be traffic generators, failure/fault injectors, interrupt generators, etc. These provide the stimuli to test and validate the models and are utilized for experimentation support.

This subsection will use the DTSS example to demonstrate and observe a typical telephone call connection within a certain amount of time as a test of basic DTSS functionality. Hence the environment will require a model of a typical telephone and user as well as a traffic generator for background telephone traffic. The user module will possess a simple behavior: lift up the handset, wait for a dial tone, dial a number, talk for a specified random time period (if the connection is made), and hang up.

The telephone module will act as the interface between the user and the switch by performing such operations as sending the dialed digits to the switch, providing the different tones to the handset (user feedback), and allowing voice messages to pass to and from the handset. A *load_generator* module is also used to initiate a request for a call to a user module at random intervals to random destinations. The structural representation of the environment model is represented in Fig. 7.17. Figure 7.18 provides a predicate/transition-net representation of the behavior for the telephone module and user modules.

DASE provides simulation support for design exploration. Once the main DSL program has been loaded into the environment, all library

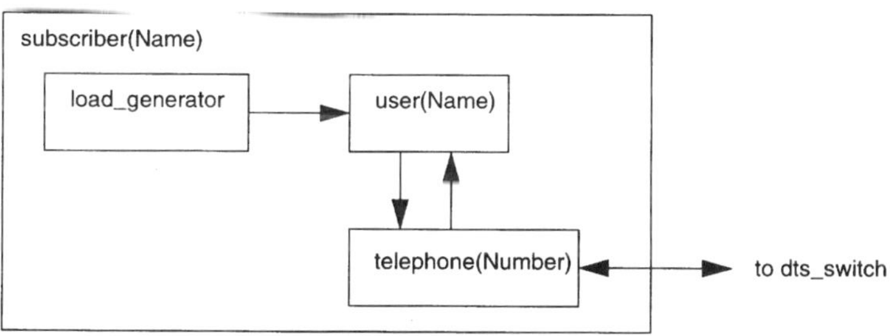

Figure 7.17 DSL representation of environment for DTSS.

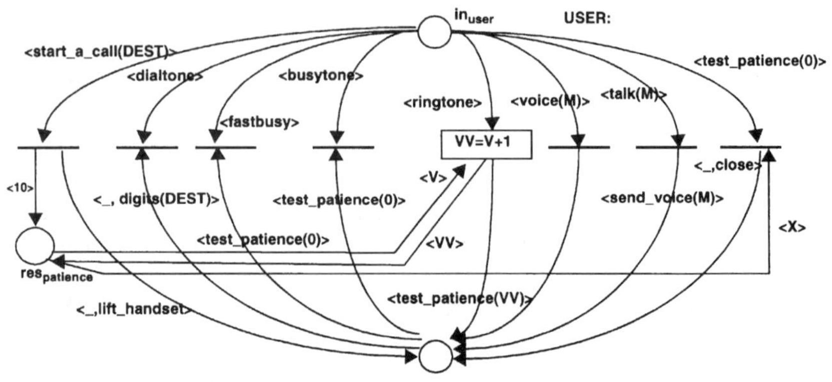

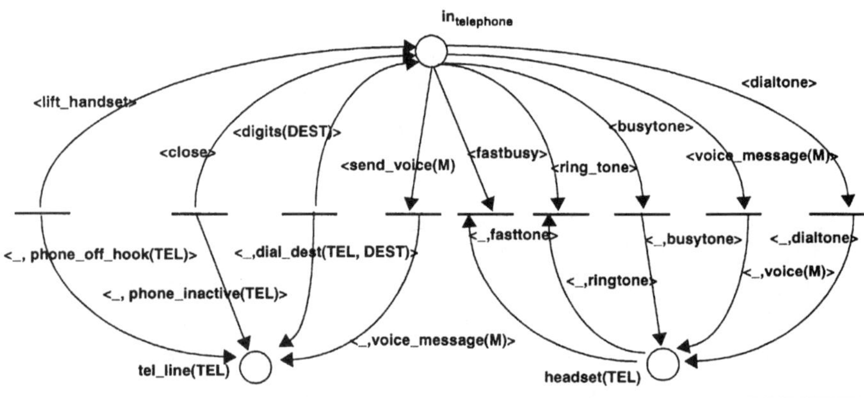

Figure 7.18 Predicate net of telephone and user.

modules are also loaded, configured, and instantiated. The DSL simulator automatically creates an internal record of the hierarchy tree. This tree is used to define the observation level for the model during simulation. The hierarchy tree for the DTSS is shown in Fig. 7.19.

By default, the observation level is set to depict the maximum detail, which corresponds to level 4 in this example. However, if the telephone call setup, which is a relatively high-level view of the system, is of interest, then the internal nuances of the switch should be hidden from the user so that there is not an overload of information. This implies setting the observation level to the first level with the statement *set_level(1)*. Based upon the hierarchy tree, this level will restrict the simulator output to only messages passing between the switch and subscribers.

The observational level can be changed at any time during a simulation, allowing the user to zoom in or out as required. The library

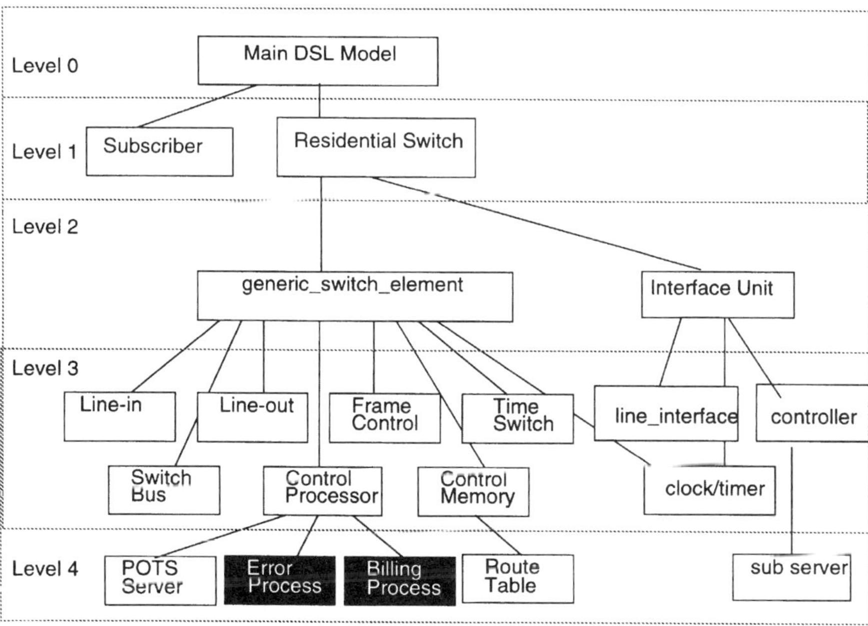

Figure 7.19 Hierarchy for DTSS.

modules may also contain associated internal constraints. For example, some telephony practices dictate a maximum call delay (time from when a call is dialed until a ring tone is heard) of 450 μs. This time is incorporated as a constraint (upper bound) for the ring-tone message to be delivered to the destination. A sample snapshot of a simulation run is shown in Fig. 7.20.

The listing traces the messages for subscriber(0), who initiates a call to subscriber(5). Part 1 of the listing shows the level 1 messages observed by the simulator. These include lifting of the handset, dial tone, dialing the destination number, and the related messages to the switch from the telephone. No other level 1 message is observed in the simulation run until 400 μs later. The output for this time period is shown in part 2. A ring tone is heard on subscriber(5)'s telephone as well as the caller's, the phone is answered, and a "hello" voice message is sent to the originator. Hence the phone call is established.

In this example, the call delay is 400 μs, and hence it is within the defined system constraint. This constraint is used to understand and set some performance bounds for a system-level parameter (or metric). Lower-level design-related parameters are also a part of the model. A significant design constraint is placed upon the control processor in the generic switch element. It must be able to process new incoming data before they are overwritten. Hence there is a con-

```
DSL SIMULATOR: Observable level set to 1

----> CURRENT SIMULATOR TIME is 0.0 microseconds.
==========================================================
--------------------------------------------------
DSL SIMULATOR: Scheduling message from subscriber(0) via subscriber(0) to telephone(0)
Source Port: hand(0) Message is: lift_handset
--------------------------------------------------
--------------------------------------------------
DSL SIMULATOR: Scheduling message from telephone(0) via residential_switch(0) to controller(0)
Source Port: tel_line(0) Message is: phone_off_hook(0)
--------------------------------------------------
--------------------------------------------------
DSL SIMULATOR: Scheduling message from telephone(0) via telephone(0) to subscriber(0)
Source Port: headset(0) Message is: dialtone
--------------------------------------------------
--------------------------------------------------
DSL SIMULATOR: Scheduling message from subscriber(0) via subscriber(0) to telephone(0)
Source Port: hand(0) Message is: digits(telno(0,1,5))
--------------------------------------------------
--------------------------------------------------
DSL SIMULATOR: Scheduling message from telephone(0) via residential_switch(0) to controller(0)
Source Port: tel_line(0) Message is: dial_dest(0,telno(0,1,5))
--------------------------------------------------
```

PART 1

```
----> CURRENT SIMULATOR TIME is 400.0 microseconds.
==========================================================
--------------------------------------------------
DSL SIMULATOR: Scheduling message from telephone(0) via telephone(0) to subscriber(0)
Source Port: headset(0) Message is: ringtone
--------------------------------------------------
--------------------------------------------------
DSL SIMULATOR: Scheduling INTERNAL message from subscriber(0)
Message is: test_patience(9)
--------------------------------------------------
--------------------------------------------------
DSL SIMULATOR: Scheduling message from telephone(5) via telephone(5) to subscriber(5)
Source Port: headset(5) Message is: ringtone
--------------------------------------------------
--------------------------------------------------
DSL SIMULATOR: Scheduling INTERNAL message from subscriber(5)
Message is: talk(hello)
--------------------------------------------------
--------------------------------------------------
DSL SIMULATOR: Scheduling message from subscriber(5) via subscriber(5) to telephone(5)
Source Port: hand(5) Message is: lift_handset
--------------------------------------------------
--------------------------------------------------
DSL SIMULATOR: Scheduling message from subscriber(5) via subscriber(5) to telephone(5)
Source Port: hand(5) Message is: send_voice(hello)
--------------------------------------------------
--------------------------------------------------
DSL SIMULATOR: Scheduling message from telephone(5) via residential_switch(0) to controller(1)
Source Port: tel_line(5) Message is: phone_off_hook(5)
--------------------------------------------------
--------------------------------------------------
DSL SIMULATOR: Scheduling message from telephone(5) via residential_switch(0) to controller(1)
Source Port: tel_line(5) Message is: voice_message(5,hello)
--------------------------------------------------
```

PART 2

Figure 7.20 Simulation snapshot.

straint imposed upon the control processor (in particular, the *frame_select* message must be processed before a certain upper bound). If the number of channels per line were increased to 32 and the number of interface units attached to the switch were increased to 20, then the demands upon the control processor would be greater. The defined control processor would not be able to keep up with the incoming traffic stream, and a message would be issued from the simulator, such as

```
DSL SIMULATOR: WARNING: Upper_limit Constraint Violation!
Module: control_processor(0) Message: frame_select(15)
Constrained Variable 0.1 exceeds upper limit of 0.051
```

Accordingly, the user can decide upon the course of the design exploration. Options include the traditional design decisions, such as replacing the control processor with a faster alternative or adding multiple processors. Alternatively, the design constraint can be relaxed or ignored to view the impact upon the overall system behavior.

As in all modeling environments, the ultimate design decisions are directed by the user. Hence the designer's modeling approach and style will effectively dictate the size of the resultant VHDL code. Some of the different modeling approaches are further illustrated in the next chapter as detailed case studies, providing system design and modeling solutions using DASE.

Chapter

8

Case Studies

This chapter will present case studies dealing with the application of DSL to different design problems. The examples are complementary to the DTSS example developed in earlier chapters. The studies are intended to present the generality and applicability of environments such as DASE within the domain of telecommunication systems. Each study focuses on different capabilities of the environment. The studies provide different levels of detail and elaboration with respect to the DSL code, presenting detailed examples of behavioral representation of design entities.

The first study will elucidate the advantages of DASE in the design of a popular technology in broadband switch design: asynchronous transfer mode (ATM). The study is intended to provide insight into the DSL modeling process as it presents a model of an ATM switch and applies it to an ATM network. Model reusability is also illustrated through the incorporation of some modules from the first case study in the ATM design. A network simulation is presented in detail in the second case study.

The second study will detail the DSL modeling of three broadcast protocols used in the support of reliable distributed computing systems. This example provides the most detail concerning behavioral descriptions within DSL. After a detailed walk-through of the design of the protocols, an example node model utilizing the protocols will be introduced. The chapter will conclude with a simulation of nodes communicating with the broadcast protocols through an ATM switching network.

ATM Switch Design

The technology push in telecommunications has blurrred the distinction between voice and data services. While most services today

TABLE 8.1 Information Rates for Data Services

Service	Burstiness	Information rate (bits/sec)
Basic voice	2	6.4×10^3
Interactive data	10	10^5–10^7
Bulk data	1–10	10^5–10^6
Standard video	2–3	10^7
High-definition TV	1–2	10^8–10^9

address voice and low-speed data, future demands will require the transparent use of these and more demanding services, such as high-definition TV and high-speed data, across a common medium. Apart from placing significant performance requirements on a system, these services have significantly different characteristics. For example, 50 percent of traditional voice traffic contains silent periods, which translates to a burstiness factor of 2 (the ratio between the maximum and the average information rate). This factor helps to determine the design requirements for a switch. Designing for the peak information rates implies wasted bandwidth at lower rates, while design for the average rates results in loss of quality in service during peak periods. The burstiness and information rates of a variety of services are tabulated in Table 8.1 (Prycker 1991). It can be observed that the demands on the system to support the different services will vary from one service to the next.

ATM has been proposed as a potential method for incorporating constant- and variable-bit-rate services such as voice and data into one transmission and switching solution. ATM is essentially a connection-oriented data-link-level protocol, with its main characteristics being summarized as follows (ATM 1993):

1. *No error control on a link basis. Hence some information may be lost. The assumption is that the transmission medium quality is good enough to keep such losses within acceptable limits.*
2. *The information field in ATM is small. Information is transferred in small fixed length cells of 53 byte lengths of which 48 are reserved for information and 5 are header information.*
3. *No flow control on a link basis. Information may be lost due to overflowing queues within a switch. This risk is minimized by statistical allocation of resources. The information fields are also relatively small in ATM so that a momentary loss of information may not be a major impact on the service.*

4. *ATM is connection oriented. Hence a call setup phase exists between the service requester (an ATM terminal) and a service provider (the ATM switch).*
5. *Header functionality and size are minimal in ATM. This allows for fast hardware switching of information once a connection has been established.*

An ATM cell for an ATM switch is shown in Fig. 8.1. The 5-byte header comprises a VPI (virtual path identifier), VCI (virtual channel identifier), PTI (payload type indicator), CLP (cell loss priority), and HEC (header error control). The VPI/VCI fields are the virtual path and channel identifiers that represent the virtual circuit that the cell is traversing. A VCI is a 16-bit address; hence each VPI can contain up to 64K VCIs. These addresses are local to the switch; hence a connection table within switches identifies unique translations between incoming and outgoing VPI/VCI pairs. The PTI and CLP are used by the network for congestion control and identifying cell priorities. If the PTI indicates a significant increase in congestion on a line, variable-bit-rate cells are given a lower priority (relative to constant-bit-rate cells). In the event of buffer overloading in the switches, the lower-rate cells are discarded. The last header field is the HEC, which provides for error control across the 4 bytes of header information.

It should be noted that for smaller ATM devices, such as ATM hubs which mulitplex a set of users across a single access point to an ATM switch, the 4 bits of the first octect are intended to be used for generic flow control between the hub and the users. At present, there is no consensus within the industry with regard to the way in which these bits are treated; hence they are omitted in this example. ATM is initially expected to operate in a permanent virtual circuit mode. This implies that virtual paths and circuits are allocated ahead of time by the service provider and that call-setup operations are not performed on a per-call basis as for traditional voice switching. As the signaling

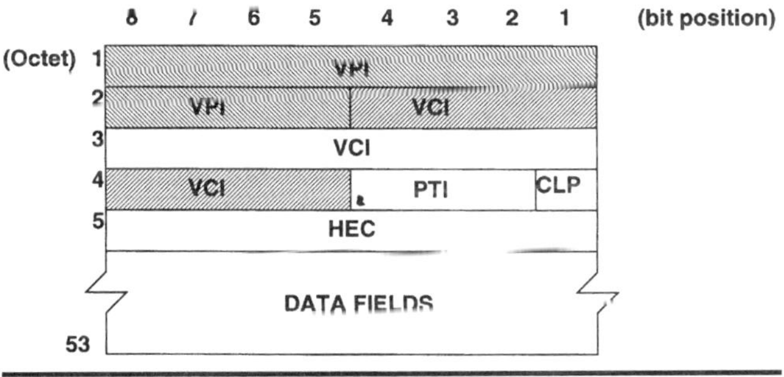

Figure 8.1 ATM cell structure.

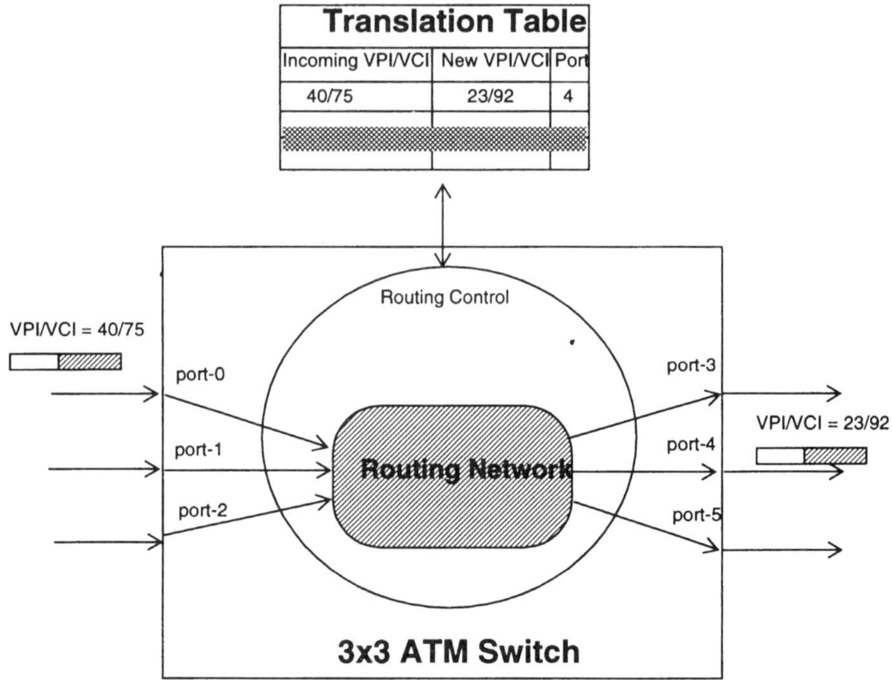

Figure 8.2 Basic ATM routing operation.

aspects for ATM become clearer, ATM services are likely to incorporate on-demand circuits (Wernik 1991).

The basic operation of an ATM switch operating in a permanent connection mode is shown in Fig. 8.2. The switch contains three input and three output ports. Definitions of the permanent circuits are contained within a translation table in the switch (these would typically be provided by the service provider or administrator). The addresses within the table are local to the switch. The VPI/VCI address field of each incoming cell is examined by the routing control in the switch and used as a reference address to the translation table, where a new VPI/VCI address is identified along with the destination output port for the cell in the switch. The figure presents an example of a cell address translation (incoming address 40/75 is translated to outgoing address 23/92). The cell is then routed (through a routing network) to the respective output port with the new VPI/VCI address until it finally reaches the destination.

There are many different design alternatives and decisions that can be considered in the design of an ATM system suitable for fast switching of ATM cells (Zegura 1993, Bae 1991). This section will present an example of a permanent-virtual-connection-oriented central queue–based design with output buffering. The example is intended to demonstrate the decreased level of modeling complexity (as compared to the digital time-space switch) as well as the applicability of DSL in futuristic telecommunications applications.

The DSL model

The ATM switch design will be introduced starting from the most abstract view, and increasing in model granularity as the lower modeling levels are revealed (as in the DTSS example). The delay parameters are omitted so as not to distract attention from the behavior. The evolution of the model is depicted in Fig. 8.3. At the topmost level within a *main* model, there is an instantiation of an ATM switch:

```
isa(atm(1), atm_switch(1,5)).
```

The above DSL statement indicates that *atm(1)* is an *atm_switch* with a parameter *5*. The parameter in this case is the size (number of input/output lines) of the switch that is modeled.

The next stage, depicted in Fig. 8.3*b*, illustrates the different functionalities within the switch. The rationale is similar to that of the DTSS design. There is a need for input and output interface modules that will

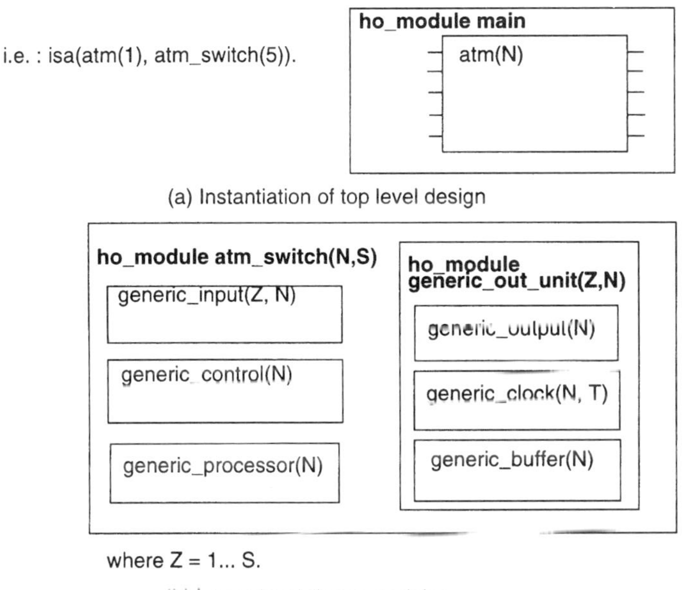

Figure 8.3 ATM switch design.

accept/transmit cells across a serial line. Since cells will be moving at relatively high rates, the design of these units should be simple and modular so that they can be synthesized readily into fast hardware. The output modules will consist of a buffer module, a clock module (which is reused from the DTS design), and a transmission module. A control unit is needed to move incoming cells from the input modules to temporary buffers, and a processor/memory module is required to manipulate the cell headers and route the cells to the appropriate output modules.

The input modules

The input modules need to capture incoming cells into a single cell latch, signal the control module (*new_cell*), and send the cell to the module when signaled back (*read_rqst*) from the control module. Intuitively, the module should have an input port (*inline*) to receive the cells and a port to interact with the control module (*control_port*). A typical description of the module in DSL is given in Fig. 8.4.

The first message that the module is capable of interpreting is the incoming cell *new_cell*. Two possible sets of behavior for *new_cell* are defined. A condition is used to check whether the cell in the buffer has been read; if not, it is discarded and a message is printed to this effect. In the *new_cell* description, a new cell is stored in *cell_buffer* and a *new_cell(INDEX)* message is sent to the module's *control_port* (*INDEX* simply identifies the input module that is signaling).

The second message that is applicable to the module is *read_rqst*. This message initiates a message *cell_info* to the *control_port*. The

```
module(generic_input(INDEX,NODE),
[
(new_cell(VPI, VCI, DATA):-
check_res(cell_buffer, (VPX, VCX, DX, unprocessed)),
    print("Lost an incoming cell at address (VPI/VCI) "),
    print(VPX), print("/"), print(VCX),
    set_res(cell_buffer, (VPI,VCI,DATA, unprocessed)),
    send(-,control_port(INDEX), new_cell(INDEX))
),
(new_cell(VPI, VCI, DATA):-
    set_res(cell_buffer, (VPI,VCI,unprocessed)),
    send(-,control_port(INDEX), new_cell(INDEX))
)
(read_rqst:-
check_res(cell_buffer, (VPX,VCX,DX,XX)),
    set_res(cell_buffer, (VPX,VCX,DX, processed),
send(-,control_port(INDEX), cell_info(VPX,VCX,DX))

)]).
```

Module Variables:
0=< INDEX < SIZE,
NODE = switch number -
the above two variables define a unique input module number

Module Resources:
name: cell_buffer.
Values: (V1, V2, DATA, FLAG)
where
V1 is the VPI address,
V2 is the VCI address,
DATA is the cell payload,
FLAG = {processed, unprocessed} indicates if data from the buffer has been successfully processed.

Figure 8.4 Input module description.

intent is to model the reading of the buffer data. The *FLAG* variable of the *cell_buffer* is also set to *processed* (the cell data have been read).

The control module

The control module will require an *input_port* to interact with the input module(s), and a port (*proc_port*) to communicate with the processor module can also be defined. The behavioral description for the control module is relatively straightforward:

```
module(generic_control(I),[
(new_cell(INPUT):-
        send(-, input_port(INPUT), read_rqst)
),
cell_info(VPI,VCI,DATA):-
        send(-,proc_port(I), new_message(VPI, VCI, DATA)) )]).
```

The module accepts a *new_cell* message (from the input modules) and replies with a *read_rqst* message to the sender. It will then accept a *cell_info* message carrying the cell information. The cell now must be sent to the processor module for correct address translation and routing. This is accomplished with the *new_message(VPI,VCI,DATA)* message.

The processor module

The address translation and routing function is performed within the processor module. Traditionally, the routing table would reside in a memory module accessible by the processor and other modules. In this example, the memory component is embedded in the processor module to emphasize that the memory/processor interface is a fast one. This example assumes permanent-virtual-connection-oriented ATM; hence call setup is assumed to be predefined. The routing information will reside within a *connect_table* resource in the processor module. This resource contains five fields: Fields 1 and 2 define the incoming VPI/VCI pair, fields 3 and 4 define the new (outgoing) VPI/VCI pair to use, and field 5 defines the address of the physical *output_unit* to which to send the cell. The behavior of the processor module is represented as

```
module(generic_processor(I),[
(new_message(VPI, VCI, DATA):-
        check_res(connect_table, (VPI, VCI, NVPI, NVCI, OUTLINE)),
        send(-, out_port(OUTLINE), atm_cell(NVPI, NVCI, DATA))
),
(new_message(V,VC,D):-
        print("ERROR: undefined circuit
connection"),print(V),print(VC))]).
```

Consequently, the processor will forward an *atm_cell* message to the appropriate *output_unit* module.

The *output_unit* module

The *output_unit* module is a higher-order module consisting of three other modules: *generic_buffer, generic_output,* and *generic_clock.* The *generic_buffer* module provides a simple FIFO queuing function for messages directed to a given *output_unit.* Consequently, the module must accept *atm_cell* messages from the processor module and maintain them in a queue for processing. The DSL behavioral code is given below:

```
module(generic_buffer(UNIT,INDEX), [
(next_cell:-
        check_res(queue, [(null, null, null) | Q])
),
(next_cell:-
        check_res(queue, [(VPI, VCI, DATA) | Q]),
        set_res(queue, Q),
        send(-, output_port(INDEX), next_info(VPI,VCI,DATA))
),
(atm_cell(VPI,VCI,DATA):-
        check_res(queue, Q),
        append([(VPI,VCI,DATA)],Q, NQ),
        set_res(queue, NQ)) ]).
```

The *queue* can be depicted in terms of a resource. The top element of the *queue* is automatically removed through a *next_cell* message from the control module. This information is passed through the use of a *next_info* message from the buffer module. To reduce communication overhead, this message is sent only if there is a message waiting in the queue (no information is represented by the null entry in the queue).

The *generic_output* module transmits ATM cells using cell data provided by either the *generic_buffer* module or a *NULL* cell contained within an internal single-cell buffer. All this is synchronized through the *generic_clock* module to generate the appropriate bit rates for cell transmission. The *generic_clock* used in the DTSS example (from the earlier chapters) is reused here. The code for the *generic_output* module is

```
module(generic_output(UNIT,INDEX),[
(clock(X):-
        check_res(out_buff, (VPI,VCI,DATA)),
        set_res(out_buff, (null,null,null)),
        send(-, out_port(INDEX), new_cell(VPI,VCI,DATA)),
        send(-, queue_port(INDEX), next_cell)
),
(next_info(VPI,VCI,DATA):-
        set_res(out_buff, (VPI, VCI, DATA))) ]).
```

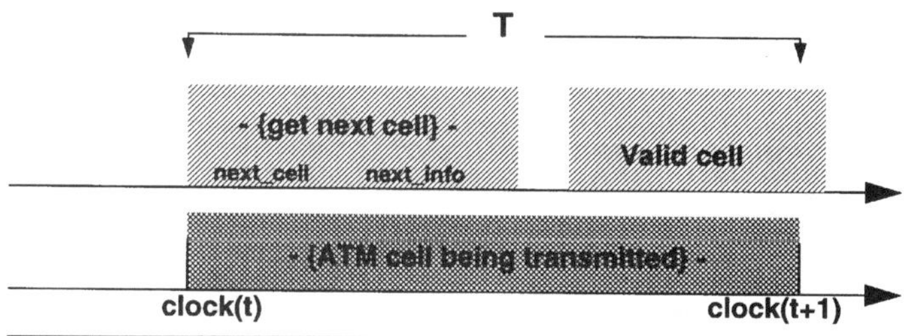

Figure 8.5 High-level ordering of messages.

The timing is such that the *generic_output* module is ready to transmit a message (*cell*) when given the *clock(X)* message from the *generic_clock* module. The typical high-level ordering of key messages is depicted in Fig. 8.5. Intuitively, an ATM cell is being transmitted between each clock message. The clock messages are provided by the same *generic_clock* module described in the time-space switch example; however, since the timing of ATM does not require frame definitions, the *frame_sync* message generated by the *generic_clock* is not used. The transmission period *T* is dependent upon the transmission rate of the line. The CCITT recommends two rates for ATM: 155Mbits/s and 622Mbits/s for medium and backbone traffic, respectively. Considering a medium traffic rate, this implies that each ATM cell takes 2.74 μs of transmission time *T*. This time is expressed within the *generic_clock* module as *Clock_rate*.

Network model—module reuse

After definition of the specific modules, the higher-order definitions provide the structural description of the underlying modules. The structural description of the switch is shown in Fig. 8.6. The figure shows only one incoming and one outgoing line. For a switch of size *N*, the *generic_input* and *generic_out_unit* modules would be replicated *N* times. The depicted switch is capable of processing messages adhering to the ATM protocol.

The defined ATM model can be reused to create a network of ATM switches capable of supporting routing and control of permanent circuits. The ATM model has already made use of the *generic_clock* module described in the DTSS example. The *generic_input* modules are also a particular instance of the DTSS's *line_in* module. Further module reuse can be achieved if voice circuit emulation over an ATM network is desired. The ATM switch would then functionally be similar to the *generic_switch_element* of the DTSS example. The ATM switch would require an additional understanding of how to establish a

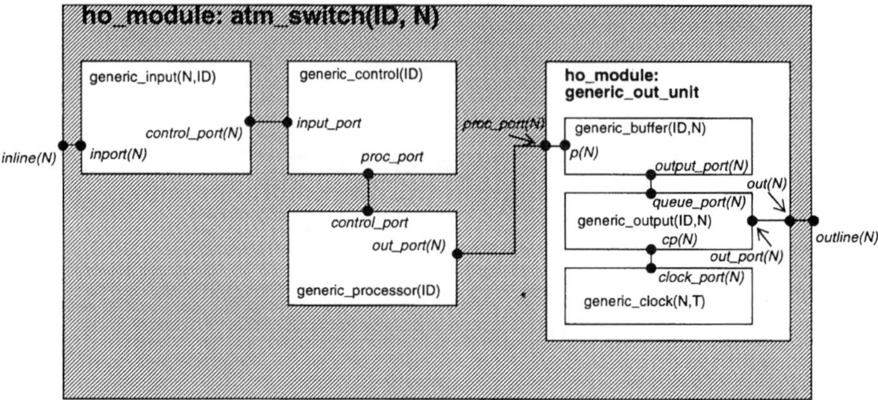

Figure 8.6 Network model of switch (*1995 IEEE*).

demanded voice circuit. The desired behavior can simply be inherited by the ATM switch's *generic_processor* from the call processing software module (*pots_server*) described as part of the DTSS example in the earlier chapters. In this case, incoming ATM cells containing voice information will be handled by the *pots_server* as in the DTSS example. Considering the lines of DSL code in the behavior of the ATM switch, approximately 65 percent of the ATM switch's behavior is composed of reused code available from the library.

A network composed of ATM switches for modeling different traffic scenarios can now be created. For example, Fig. 8.7 shows a network of four switches in a particular configuration. Figure 8.8 lists the code that is required for defining the network interconnections.

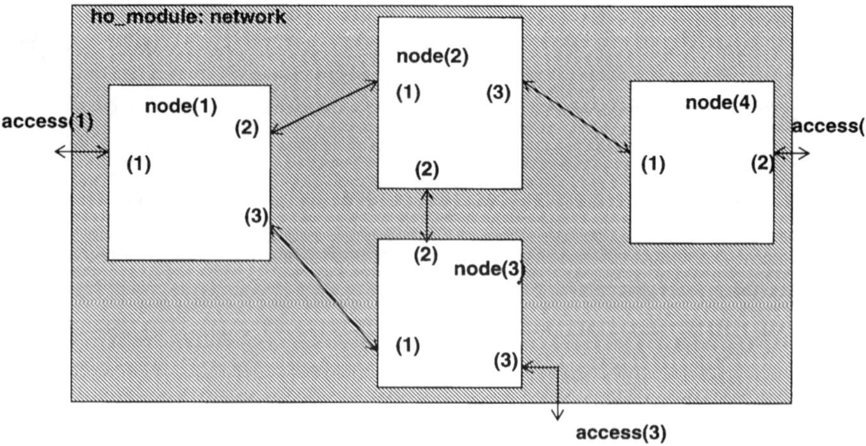

Figure 8.7 Network example.

```
sa(node(1), atm_switch(1,3))
isa(node(2), atm_switch(2,3))
isa(node(3), atm_switch(3,3))
isa(node(4), atm_switch(4,2))

path(node(1), node(2), [outline(2), inline(1)])
path(node(2), node(1), [outline(1), inline(2)])
path(node(1), node(3), [outline(3), inline(1)])
path(node(3), node(1), [outline(1), inline(3)])
path(node(2), node(3), [outline(2), inline(2)])
path(node(3), node(2), [outline(2), inline(2)])
path(node(2), node(4), [outline(3), inline(1)])
path(node(4), node(2), [outline(1), inline(3)])

path(node(1), network, [outline(1), access(1)])
path(node(3), network, [outline(3), access(3)])
path(node(4), network, [outline(2), access(2)])
path(network, node(1), [access(1), inline(1)])
path(network, node(3), [access(3), inline(3)])
path(network, node(4), [access(2), inline(2)])
```

Figure 8.8 DSL description for defining connections.

The underlying network model is relatively complete. The only thing that is required is the routing (permanent virtual path connections) definitions between nodes. This is established by defining the respective *connect_table* resource in each node's processor module. Depending upon the assertions upon these resources, all possible combinations of virtual paths and circuits for the given network can be described.

A sample routing setup is described in Table 8.2, where the entries for each node's connect_table are given. The defined setup describes

TABLE 8.2 Sample Routing Table

Node #	Incoming VPI/VCI	Outgoing VPI/VCI	Output port #
1	1/1	2/1	2
1	1/2	3/1	3
1	4/1	1/1	1
1	5/1	1/1	1
2	1/X	4/X	1
2	2/X	6/X	3
2	8/X	7/X	3
3	3/1	4/1	3
3	1/1	8/1	2
4	6/1	2/1	2
4	2/1	1/1	1
4	7/1	1/2	2

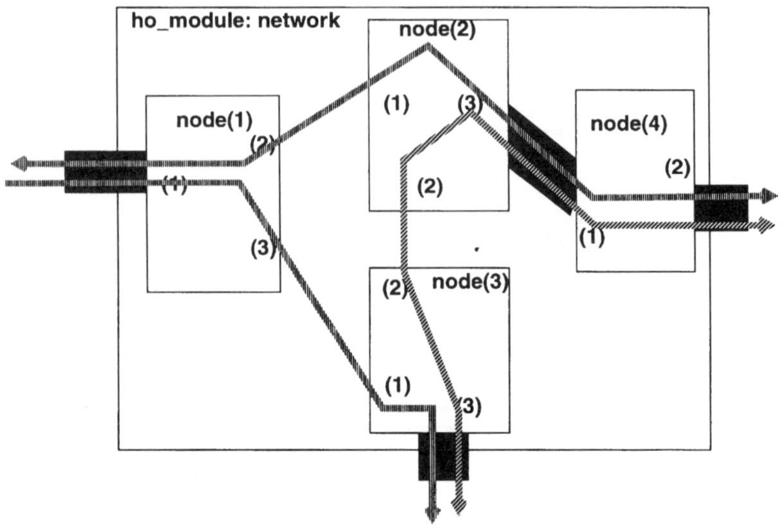

Figure 8.9 Connections on the ATM network.

virtual connections between the nodes as shown in Fig. 8.9. Every VPI/VCI address of an incoming cell is translated to a defined outgoing one at each node. The sample connection description defines three virtual connections. More connections can be defined by the user beforehand, using library configuration rules, or at run time, using DSL *set_res* or *create* commands.

As in the digital time-space example, this model can be exercised using random traffic or traffic generation modules (such as a user or a telephone). The ATM network model will be incorporated with the protocol models described in the next section to demonstrate a complex model of broadcast protocol executing over an ATM network.

Reliable Distributed Broadcast Protocols

Protocols provide the basic formalism to support communication between computing agents in a distributed computing environment; hence protocol modeling is essential to the study of computer communication systems. This section presents a case study of the modeling of a set of broadcast protocols for reliable communication within a distributed environment. Typically an application will utilize several layers of protocols to communicate; hence it is generally desirable to identify different layers of functionality in models as well (Bochmann 1990). The protocols to be modeled in this section (ABCAST, CBCAST, and GBCAST) (Birman 1993) are assumed to reside between higher- and lower-level services. Although these services are not modeled in

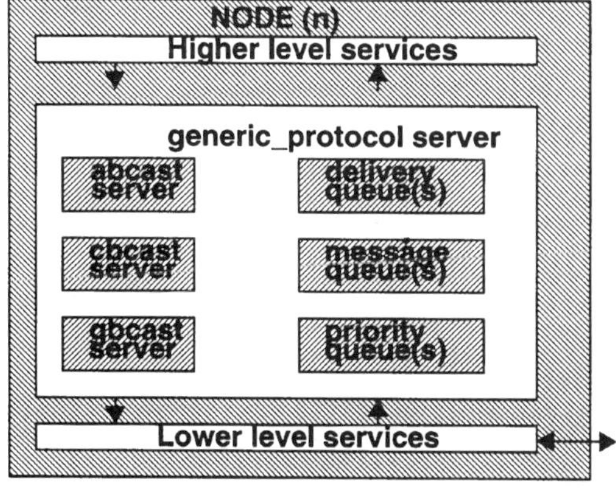

Figure 8.10 Processing node representation.

detail, they provide and request the necessary services from the three protocols. The structure of a processing node using these protocols is shown in Fig. 8.10. The *generic_protocol* server directs the upper- and lower-layer requests to the respective protocols. The upper layer sends requests for communication using one of the three broadcast protocols, whereas the lower layer provides the actual transmission-level facilities, such as send request and acknowledgments.

The section will first describe DSL models of the three protocols used to ensure reliable distributed broadcasts. The protocols ABCAST, CBCAST, and GBCAST are described in further detail in Birman (1987). The model of a node using the structure presented in Fig. 8.10 will be described, with the lower-level services tailored to interface with the ATM protocol presented in the previous section.

The ABCAST protocol

The ABCAST protocol is one of three broadcast process primitives (within a distributed environment) and ensures that the order of a broadcast message received at multiple destinations from a source process is the same even though the order is not predetermined. The DSL model describes node modules, which represent distributed sites. A high-level program (*high_level_services*) will generate processes that request ABCAST communication with other nodes through a *low_level_services* module. To support the ABCAST primitive, each node will have an *abcast_server*, a *priority_queue*, and a *delivery_queue*. The structural representation of this model is given in Fig. 8.11. The algorithm presented in Birman (1987) for ABCAST is presented below with the DSL representation.

The algorithm is composed of four steps.

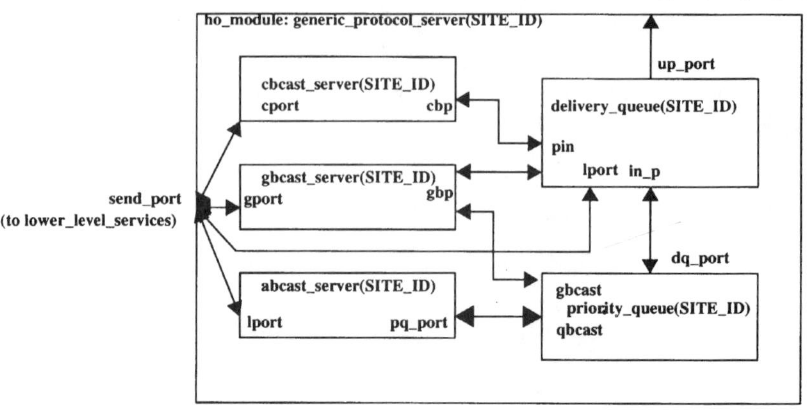

Figure 8.11 ABCAST model in DSL.

Step 1. *The sender transmits a message (Msg) with a unique label to its destinations.*

This is represented through a *form(abcast, PROC_ID, Dest, Msg)* message sent to the *abcast_server* (from the upper-level protocols). *PROC_ID* is the source process identifier, *Dest* is a list of destinations, and *Msg* is the message to be transmitted. This segment of DSL code is

```
(form(abcast, PROC_ID, Dest, Msg):-
        check_res(label_count, COUNT),
        NCount is COUNT+1,
        list_length(Dest, No),
        delay(X),
        set_res(label_count, NCount),
        create(NCount, (Dest, No, 0)),
        send(_, lport(SITE_ID), msg_rqst(Dest, abcast(Dest, Msg,
NCount, PROC_ID,
SITE_ID))))
```

Two resources are utilized above: *label_count* and *NCount.* The first is used to generate unique labels for the messages, and the latter is used to keep track of how many destinations within a process group have responded during step 2 of the protocol. The last line will send the abcast message to a lower-level protocol handler (connected through port *lprot(SITE_ID)).*

Step 2. *Recipients add Msg to a priority queue associated with label, marking it as undeliverable. A priority is assigned to the message (NPri) larger than the largest in the queue, with a process identification (PROC_ID) added as a suffix. This suggested priority is then transmitted to the sender.*

The DSL code given below depicts the behavior for this step. However, an interaction with the *priority_queue* module (connected through the *pq_port(SITE_ID)*) is not shown here, but will be presented later in the section. The *add_new* message sent to the *priority_queue* module requests that the module add the new ABCAST message to the priority queue.

```
(abcast(Msg, Label, PROC_ID, Sender_ID):-
        send(_, pq_port(SITE_ID), add_new(abcast, Sender_ID, Msg,
        Label, PROC_ID)))
```

The *priority_queue* module will consequently send an *updated_priority* message (shown below) that provides the suggested priority (*NPri*). This value is sent back to the sender as shown below:

```
(updated_priority(abcast, SITE, Label, NPri):-
        send(_, lprot(SITE_ID), msg_rqst(SITE, abcast(SITE, sug
        priority (NPri), Label))))
```

Step 3. *The sender waits for all the suggested priorities from the destinations, computes the maximum of all the values, and sends this value to all the destinations, thus guaranteeing correct order of delivery.*

The DSL representation given below is used to compute the maximum priorities as well as to check whether all responses have been collected from the destinations. The largest priority value is maintained within the resource *Label* (*NCount* in step 1), along with a flag (*State*) indicating whether all replies have been received.

```
(abcast(sug_priority(Value), Label):-
        check_res(Label, (Dest, State, Prior)),
        NState is State -1,
        set_res(Label, (Dest, NState, Prior)),
        Value>Prior,
        set_res(Label, (Dest, NState, Value )),
        send(check(Label))),
(abcast(sug_priority(Value), Label):-
        send(check(Label)),)

(check(Label):-
        check_res(Label, (Dest, State, Prior)),
        State>0),
(check(Label):-
        check_res(Label, (Dest, State, Prior)),
        delay(1),
        remove(Label, X),
        send(_, lport(SITE_ID), msg_rqst(DEST, abcast
        (set_priority(Prior), Label))))
```

Step 4. *The destinations update the priority for Msg to the new value, mark the message as deliverable, and re-sort the priority* *queues. The destinations then move messages in order of increasing priority from the priority queues to a delivery queue. This continues as long as the priority queue remains nonempty and there is a deliverable message at the top of the queue.*

```
(abcast(set_priority(Value), Label, Proc_ID):-
        send(_, pq_port(ID), change_priority(abcast, Value, Label,
        Proc_ID)))
```

This DSL code describes the final action that is required by the *abcast_server.* A message is sent to the *priority_queue* to change the priority of the message identified by *Label.* The *priority_queue* module requires some behavioral code to communicate with the *abcast_server* and the *delivery_queue.* The *add_new* behavior is used to add a new message to a priority queue (*p_queue(PROC_ID)*). There are two cases defined.

The first case assumes that the priority queue for a process exists. Hence a new priority is assigned to the new message, and another data structure, *waiting_message(PROC_ID),* is created to keep a copy of the message and relevant information such as state (*deliverable* or *undeliverable*), message label, data, and priority value. The second case takes into account the situation in which a queue does not exist yet for a process *PROC_ID.* Hence a queue is created to hold ordered priorities, and all other operations are performed as in the first case.

```
(add_new(Protocol, Sender, Msg, Label, PROC_ID):-
        check_res(p_queue(PROC_ID), [Pri|Rest]),
        NPri is Pri+1,
        create(waiting_message(PROC_ID), (Npri, Msg, Label,
        undeliverable)),
        check_res(p_queue(PROC_ID), QUEUE),
        set_res(p_queue(PROC_ID), [Npri|QUEUE]),
        send(_, proc_port(SITE_ID), updated_priority(Protocol,
        Sender, Label, NPri))),
(add_new(Protocol, Sender, Msg, Label, PROC_ID):-
        NEW is SITE_ID/100,
        delay(1),
        create(p_queue(PROC_ID), [NEW]),
        create(waiting_message(PROC_ID), (NEW, Msg, Label,
        undeliverable)),
        send(_, proc_port(SITE_ID), updated_priority(Protocol,
        Sender, Label, New))),
```

The second behavior required by the *priority_queue* is *change_priority.* The description of this is given in Fig. 8.12. Upon receipt of this message, the module updates the priority (*Value*) of the specified message (*Label*). An internal message *check_pq* is generated to clean up empty queues and test whether the topmost element of the *p_queue* is deliverable. The latter is accomplished through another internal mes-

```
(change_priority(abcast, Value, Label, PROC_ID):-
    check_res(waiting_message(PROC_ID), (Pri, M, Label, State)),
    check_res(p_queue(PROC_ID), QUEUE),
    delay(1),
    remove_element(QUEUE, Pri, NQ),
    append([Value], NQ, Result),
    iqsort( Result, Sorted),
    set_res(p_queue(PROC_ID), Sorted),
    set_res(waiting_message(PROC_ID), (Value, M, Label, deliverable))
    send(check_pq(abcast, PROC_ID, Label)),
(check_pq(abcast, ID, Label):-
    check_res(p_queue(ID), Queue),
    last_list(Queue, TOP),
    check_res(waiting_message(ID), (TOP, M, Label, STATUS)),
    send(delivery(Label, STATUS))),
(check_pq(abcast, ID, Label):-
    remove(p_queue(ID), []),
    write('Priority Queues for ID '), write(ID), write(' freed'),nl),
(delivery(Label, deliverable):-
    check_res(waiting_message(ID), (Val, M, Label, deliverable)),
    remove(waiting_message(ID), (Val, M, Label, deliverable)),
    check_res(p_queue(ID), Q),
    remove_element(  Q, Val, NQ),
    set_res(p_queue(ID), NQ),
    send(_,dq_port(SITE_ID),delivery_of_new_message(ID, M, Label)),
    send(check_pq(abcast, ID, Label))),
(delivery(L, undeliverable):-
    write('Undeliverable message still waiting for processing.'),nl)]).
```

Figure 8.12 DSL description of priority queue.

sage (*delivery*) which removes deliverable messages from the top of the queues and sends them to the *delivery_queue* module.

The DSL description for the *delivery_queue* simply records the sequence of messages delivered to a node. The queue functionality is implicit in the manner in which modules process incoming messages. The DSL representation outputs the message identifiers as they are delivered:

```
module(delivery_queue(SITE_ID),[
(delivery_of_new_message(ID, M, LABEL):-
```

```
write(`Message processed at node: `),
write(SITE_ID),nl, write(`ID= `), write(ID),nl,
write(`label= `), write(LABEL),nl)]).
```

This is a very simple view of the delivery queue; however, the model will be refined later in the GBCAST protocol section to reflect the needs of all three protocols.

The behavior of the *abcast_server* module is described in terms of its equivalent predicate/transition net, given in Fig. 8.13. The protocol functionality is separated into two sets for clarity: the sender and receiver portions. Each handles the respective functionality of the protocol. Input places (identified by the darkened circles) will allow their respective token types (such as *form_abcast(D,M,P)*). These tokens arrive from other subnets (representing the other modules). A controlling net (omitted here for the sake of clarity) ensures that the specified incoming tokens are directed to the appropriate input places as well as ensuring safeness within each subnet. Operations on transitions are indicated within “<>,” while predicates are labeled on arcs. By convention, a label beginning with a capital letter is an unassigned variable.

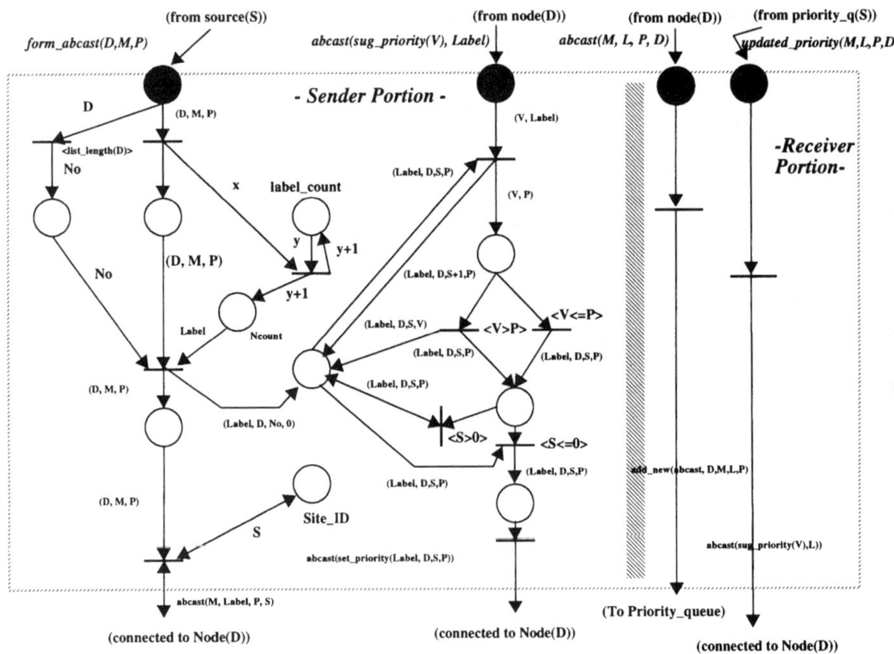

Figure 8.13 Predicate net representation of ABCAST.

Even at this level of representation, some design partitioning is being assumed. For example, the protocol and priority queue management aspects are separated into different modules, which implies that the eventual design would also favor such a distinction.

The CBCAST protocol

The CBCAST protocol is less restrictive than ABCAST, while preserving the causality of messages. It enforces a minimally synchronized delivery order. In this case, if *a* and *b* are two concurrent processes, CBCAST may deliver the processes in different orders, whereas with ABCAST, an order would be agreed upon by the destinations. The perceived advantage over ABCAST is the performance of the protocol. CBCAST does not delay transmission of messages and can therefore perform significantly faster.

The algorithm for an implementation of the CBCAST protocol as presented by Birman is as follows:

> *Each process p has a message buffer BUFp which contains copies of messages sent to and from it. A message B is transmitted from BUFp at site s to BUFq at site t as follows:*
>
> 1. *When a CBCAST message B is initiated by p, an ID(B) is associated with the message and the list of destinations are added to REM_DESTS(B) to keep a record of destinations that are destined to receive the message.*
> 2. *A transfer packet <B1, B2,... > is created which includes all messages B´ in BUFp such that B´ must precede B and REM_DESTS(B´) is nonempty. The messages are maintained in an order of precedence to preserve causality (B1 must precede B2).*
> 3. *The transfer packet is sent from site s to t.*
> 4. *For each Bi, the destination q is deleted from REM_DESTS(Bi).*
> 5. *On the receiving site t, for each message in preceding order:*
>   5.1 *The ID(Bi) is checked. If it is associated with a message in BUFq then Bi is a duplicate and is discarded.*
>   5.2 *If q is a member of REM_DESTS(Bi), q is removed from REM_DEST(Bi), Bi is placed on the delivery queue for q and a copy of Bi is placed in BUFq.*
>   5.3 *If none of the above two cases apply, Bi is a message in transit to another process and is simply placed in BUFq.*

The CBCAST algorithm is modeled within DSL using the following resources: *buff(PROC_ID)* represents the BUFFp (where *p* = *PROC_ID*) in the algorithm above, and the variables within this resource are *(ID, M, REM_DESTS)*. *ID* is a unique identification of the message *M*, and *REM_DESTS*, as above, represents the (remaining) destinations that the message is to be sent to.

The DSL model follows the same steps. There are two possible messages that can be generated to the *cbcast_server* from a higher-level protocol: (1) a message to transmit any outstanding messages in the message buffer *buff(PROC_ID),* and (2) a request for the transmission of a new CBCAST message. Since the former is a special case of the latter, the DSL code deals with the latter. A message from a higher-level protocol is given as *form(cbcast, PROC_ID, DESTS, MESSAGE)* to create a new entry in the buffer for the process (*PROC_ID*). The former is represented by a *form_cbcast* message. This message looks at any entries in the buffer(s) where there are messages awaiting transmission, creates a transfer packet (*PACKET*), and sends the result to a lower-layer protocol for transmission. The DSL code to handle the CBCAST transmission is given in Fig. 8.14.

The DSL model also requires knowledge of how an incoming CBCAST message is to be handled (step 5 of the algorithm for the protocol). A lower-layer protocol is assumed to deliver a CBCAST message as *cbcast_message(PROC_ID, PACKET),* where *PACKET* is an ordered precedence list of messages sent to this site and *PROC_ID* is the destined process identification. The DSL behavioral code and the equivalent steps of Birman's algorithm are shown in Fig. 8.15. Step 5.0 is an additional case that is the stopping condition for the algorithm.

```
(form(cbcast, PROC_ID, DESTS, Message):-
    check_res(message_label(PROC_ID), Label),
    ID is Label+1,
    set_res(message_label(PROC_ID), ID),
    create(buff(PROC_ID), (ID, [DESTS], Message)),
    send(form_cbcast)),
(form_cbcast:-
    check_res(buff(PROC_ID), (ID, [DEST|REST], M)),
    set_res(buff(PROC_ID), (ID, [REST], M)),
    send(create_packet(PROC_ID, DEST, (ID, [REST], M) ))),


(create_packet(PROC_ID, DEST, PACKET):-
    check_res(buff(PROC_ID), (ID, REM, M)),
    not(member((ID,REM,M), PACKET))
    send(create_packet(PROC_ID, DEST, [PACKET | (ID,REM,M)])),
(create_packet(PROC_ID, DEST, PACKET):-
    send(-, proto_port, msg_rqst(DEST, cbcast_packet(DEST, PACKET))),
```

Figure 8.14 DSL description of CBCAST.

```
(cbcast_message(PROC_ID, [ ]):-                    (step 5.0)
    write('CBCAST messages all processed at '),
    write(PROC_ID),nl),
(cbcast_message(PROC_ID, [(ID, REM, M)|Rest]):-          (step 5.1)
    check_res(buff(PROC_ID), (ID, _,_) ),
    write('CBCAST message id '),write(ID),
    write(' already in buffer.'),nl,
    send(cbcast_message(PROC_ID, Rest))),
(cbast_message(PROC_ID, [(ID, REM, M)|Rest]):-          (step 5.2)
    member(PROC_ID, REM),
    remove_element(REM, PROC_ID, NEW_REM),
    create(buff(PROC_ID), (ID, NEW_REM, M)),
    send(delivery_queue(PROC_ID),_,delivery_of new message(ID, M)),
    send(cbcast_message(PROC_ID, Rest))),
(cbcast_packet(PROC_ID, [(ID, REM, M)|Rest]):-          (step 5.3)
    not(member(PROC_ID, REM)),
    create(buff(PROC_ID), (ID, REM, M)),
    send(cbcast_message(PROC_ID, Rest)))]).
```

Figure 8.15 DSL implementation of protocol.

The GBCAST protocol

The last of the three protocols to be described provides a means of communication between changing process groups. The previous two protocols assume that all the destinations are known; however, GBCAST permits communication to a group of processes whose membership may change over the course of time. To manage the group view, the protocol provides a totally ordered delivery scheme with respect to the other two protocols. The GBCAST protocol presented by Birman is outlined next.

Step 1. *The first step of the protocol orders GBCASTs with respect to ABCAST messages. A GBCAST message is composed of a destination group G and an action to be performed by the group members (action). A process p distributes the message to the member processes of group G.*

The *form* message is interpreted by the *gbcast_server* as a request to send a GBCAST message (*ACTION*) to a *GROUP*. A local view of groups is maintained within a *site_group_table* (which records which groups a node belongs to) and a *proc_group_table* (which tracks the processes' membership in groups). A resource *label_count* is also used to generate increasing unique label values for messages. The originator of the GBCAST also tracks the replies from destination members using the resource *LABEL* (line 1.9).

```
1.1  (form(gbcast, PROC_ID, GROUP, ACTION):-
1.2      check_res(label_count, COUNT).
1.3      check_res(proc_group_table, (GROUP, PROCS)),
1.4      check_res(site_group_table, (GROUP, SITE)),
1.5      LABEL is COUNT+1,
1.6      list_length(GROUP, No),
1.7      delay(DELAY1),
1.8      set_res(label_count, LABEL),
1.9      create(LABEL, (GROUP, No, 0)),
1.10     send(_, gport, msg_rqst(SITE, gbcast(SITE_ID, ACTION,
         LABEL, GROUP)))),
```

Step 2. *Following the steps of the ABCAST procotol, the recipient q places the message (tagged as undeliverable) on all ABCAST priority queues. A priority value is assigned to all the messages based on the next highest value of any message on the ABCAST queues, and the value is sent back to process p.*

```
(gbcast(Sender, ACTION, LABEL, G):-
        check_res(proc_group_table, (G, PROC)),
        create(label_member, (LABEL, PROC, Sender)),
        send(gbcast_m(Sender, ACTION, LABEL, PROC))),


(gbcast_m(Sender, Msg, Label, [])),
(gbcast_m(Sender, Msg, Label, [PROC_ID|REST]):-
        send(_,pq_port(SITE_ID),add_new(gbcast, Sender, Msg, Label,
        PROC_ID),
        send(gbcast_m(Sender, Msg, Label, REST))),


(updated_priority(gbcast, Sender, Label, NPri):-
        send(_, gport,msg_rqst(Sender, gbcast(sug_priority(NPri),
        Label))))
```

Step 3. *After obtaining all values from group members, p sends the maximum value of all the values to the group G. The group members accordingly assign the new value and re-sort the priority queues. However, the messages are not tagged as deliverable, as in the ABCAST case. When the GBCAST messages arrive at the head of their respective priority queues, further delivery from the queues will be suspended. When the GBCAST message reaches the head of all the ABCAST priority queues, the next steps are undertaken to order the GBCASTs relative to CBCASTs.* The DSL description of this is shown in Fig. 8.16.

The DSL description presented in the figure is virtually identical to step 3 for ABCAST and is given for the sake of completeness. The dif-

```
(gbcast(sug_priority(Value), Label):-
    check_res(Label, (Dest, State, Prior)),
    NState is State -1,
    set_res(Label, (Dest, NState, Prior)),
    Value> Prior,
    set_res(Label, (Dest, NState , Value )),
    send(check(Label))),
(gbcast(sug_priority(Value), Label):-
    send(check(Label)),
    write('Priority that was received was not high enough'),nl),
(check(Label):-
    check_res(Label, (Dest, State, Prior)),
    State>0,
    write('Have not received all replies yet'),nl),
(check(Label):-
    check_res(Label, (Dest, State, Prior)),
    delay(1),
    remove(Label, X),
    check_res(site_group_table, (GROUP, SITE)),
    send(_, gport, msg_rqst(SITE, gbcast(set_priority(Prior), Label,GROUP)))),
```

Figure 8.16 DSL description of GBCAST.

ference is in the way in which the destinations are specified. In ABCAST, the destinations were given directly as a part of the protocol. For GBCAST, the site and process group tables are used to identify the destinations within a group relevant to a particular site.

Since GBCAST does not completely follow the ABCAST step 4, there are some additional differences to be considered. Unlike ABCAST, messages are not tagged deliverable at this stage. The protocol defined below will tag them as *gbcast_wait,* indicating that the message is to be delivered according to GBCAST. An additional resource, *gb_ab_order,* which keeps track of how many GBCAST members within the node still have not reached the top of their respective priority queues is maintained in the *gbcast_server.* The ABCAST step 4 represented in DSL as applied to GBCAST would then be

```
(gbcast(set_priority(Value), Label, G):-
        check_res(proc_group_table, (G, PROCS)),
        create(gb_ab_order, (G, Label, PROCS)),
        send(gbcast(set_priority(Value), Label, G, PROCS)
(gbcast(set_priority(Value), Label, G, []):-
```

```
        check_res(proc_group_table, (G, PROCS)),
        send(_,pq_port(SITE_ID),check_pq(gbcast, PROCS, Label))),
(gbcast(set_priority(Value), Label, G, [ID\REST]):-
        send(_,pq_port(SITE_ID), change_priority(gbcast, Value,
        Label, ID)),
        send(gbcast(set_priority(Value), Label, G, REST))
```

The *priority_queue* will require some additions to add the GBCAST message. This is conveniently accomplished through the *change_priority* message. The first parameter of the message shown below identifies it as a GBCAST message; hence it is tagged as being in a *gbcast_wait* state.

```
(change_priority(gbcast, Value, Label, PROC_ID):-
        check_res(waiting_message(PROC_ID), (Pri, M, Label, State)),
        check_res(p_queue(PROC_ID), QUEUE),
        delay(1),
        remove_element(QUEUE, Pri, NQ),
        append([Value], NQ, Result),
        iqsort( Result, Sorted),
        set_res(p_queue(PROC_ID), Sorted),
        set_res(waiting_message(PROC_ID), (Value, M, Label,
        gbcast_wait))),
        send(check_pq(abcast, PROC_ID, Label))
```

In addition, the priority queue will inform the *gbcast_server* when a message with a state of *gbcast_wait* is detected at the head of a *priority_queue*. This is described through the following delivery message, where a message *gbcast_wait* is sent to the *gbcast_server*.

```
(delivery(Label, gbcast_wait):-
        check_res(waiting_message(PROC_ID),
        (V,M,Label,gbcast_wait)),
        send(_, proc_port(SITE_ID), gbcast_wait
        (PROC_ID, Label))).
```

The *gbcast_server* module must process the *gbcast_wait(PROC_ID, Label)* message generated above. The description of the module is given in Fig. 8.17. The module identifies and removes the process identification (*PROC_ID*) from the appropriate *gb_ab_order* resource. When the third variable within the resource is an empty list, this implies that all members of the group are at the top of their respective priority queues (these tests are performed within the additional *test_ab* behaviors given below). At this point, a *set_wait_queue* message is sent to each process's delivery queue through the *instant_queue* behaviors.

Step 4. *A wait queue is used by the recipient processes to temporarily hold messages that would have been placed on the delivery queue by the CBCAST protocol. A list (IDLIST) is maintained which contains*

```
gbcast_wait(PROC_ID, Label):-
    check_res(gb_ab_order, (G, Label, PROCS)),
    remove_element(PROCS, PROC_ID, NEW_PROCS),
    set_res(gb_ab_order, (G,Label, NEW_PROCS)),
    send(test_ab(G, Label, NEW_PROCS))

test_ab(G,Label,[]):-
    check_res(proc_group_table, (G, PROCS)),
    check_res(label_member, (Label, P, Sender)),
    send(instant_queue(Sender, PROCS, PROCS)),
test_ab(G,Label,[]):- print(' Not all members are at head of queues'),

instant_queue(S, G, []):-print('All wait queues generated'),
instant_queue(Sender, PROC, [MEMBER|REST]):-
    send(_,_,set_wait_queue(MEMBER)),
        send(_,_,flush_cbcast(Sender, PROC, MEMBER))
    send(instant_queue(REST))
```

Figure 8.17 DSL description of CBCAST server.

IDs of all CBCAST messages that were placed on a delivery queue of the recipient processes.

The requirements for this step of the algorithm dictate the need for a more sophisticated *delivery_queue* module. The module will contain resources that define the delivery queue (*d_queue*) and the wait queue functions (*wait_queue*), as well as *idlist(ID)* to store the message labels within a *d_queue(ID)*. The DSL description also identifies the behavior of the *set_wait_queue* message sent from the *gbcast_server*. The *delivery_of_new_message* will place the message onto a *wait_queue* if it exists or directly onto the *d_queue*. In the latter case, a new message notification is sent to the upper-layer protocol.

```
(set_wait_queue(ID):-
        create(wait_queue(ID), []) )


(delivery_of_new_message(ID, M, LABEL):-
        check_res(wait_queue(ID), REST),
        set_res(wait_queue(ID), [(M, LABEL)]),
(delivery_of_new_message(ID, M, LABEL):-
        check_res(d_queue(ID), CONTENTS),
        set_res(d_queue(ID), [(M, LABEL)|CONTENTS]),
        check_res(idlist(ID), REST),
        set_res(idlist(ID), [LABEL|REST]),
        send(_,up_port(SITE_ID), new_msg)),
```

Step 5. *For a process q, if any message B in IDLIST is in BUFFq and B includes destinations that are a part of the group G, these messages are scheduled for transmission to the destinations. When all such messages are sent, q sends its IDLIST to the originating process p.*

This step requires the interaction of the *gbcast_server, delivery_queue,* and *cbcast_server* modules. First, the *gbcast_server* must generate a *flush_cbcast* message (from step 3) to the *delivery_queue.* This signals that GBCAST ordering with respect to ABCASTs is done, the GBCAST ordering with respect to CBCASTs must now be performed. The *delivery_queue* module will use this message to transfer the *idlist* resource data to the *cbcast_server* with the *gb_request* message. The *cbcast_server* will then send any messages residing in its buffer belonging to a destination that is a member of a given group.

The DSL code below is a part of the *delivery_queue* behavioral description. As a response to the *gbcast_server*'s *flush_cbcast* message, the module transmits a *gb_request* message to the *cbcast_server,* which is a request to process any messages that are to be transmitted to group members. A *cbcast_list* message which is used in step 6 of the protocol is sent to the *gbcast_server.*

```
flush_cbcast(Sender, PROCS, PROC_ID):-
        check_res(idlist(PROC_ID), IDLIST),
        send(cbcast_server(SITE_ID),_, gb_request(PROCS, IDLIST)),
        send(_, lport, msg_rqst(Sender, cbcast_list(IDLIST, PROCS,
        PROC_ID)))
```

The code given below is located within the *cbcast_server* and is a description of the actions to be undertaken upon receiving the *gb_request* from the *delivery_queue.* Essentially, each member of the REM destination list within the *buff* resource is compared with each member of the IDLIST within the *test_membership* behaviors. A *create_packet* message is subsequently generated for CBCASTs to be sent to group members (this was described in the CBCAST protocol section).

```
gb_request(P, []):- print(`All outstanding CBCASTs sent'),
gb_request(PROCS, [B\ IDLIST]):-
       check_res(buff(PROC_ID), (B, REM, M)),
       send(test_membership(REM, PROCS, B, PROC_ID),
       send(gb_request(PROCS, [IDLIST] ))
(test_membership([], G,ID,P)),
(test_membership([R1REST], GROUP,ID, PROC_ID):-
     member(R, GROUP),
     check_res(buff(PROC_ID), (ID, LIST, M)),
     remove_element(R, LIST, NEWLIST),
     set_res(buff(PROC_ID), (ID, NEWLIST, M)),
     send(create_packet(PROC_ID, DEST, (ID, NEWLIST, M)))
     send(test_membership(REST, GROUP, ID, PROC_ID))),
(test_membership([R\ REST], GROUP,ID, PROC_ID):-
     send(test_membership(REST, GROUP, ID, PROC_ID)))
```

Step 6. *After receiving all the IDLISTs, p merges the lists into a before list, which it sends to G.*

The *cbcast_list* message that was sent from all the participating group members is processed within the GBCAST originator's *gbcast_server* module. A resource *before_list(Group)* which keeps the merged version of the IDLIST as well as a list of members that have sent their lists is maintained for the group. Upon receipt of an IDLIST from a group member, the member is removed from the *before_list(Group)*. A *test_before* behavior is used to determine if all the members have sent their lists (this would be an empty list for the second variable of the before list). If so, the before list is sent to members as the message *before_list(IDlist, Group)*.

```
(cbcast_list(IDLIST, Group, PROC_ID):-
        check_res(before_list(Group), (LIST, PROCS)),
        remove_element(PROCS, PROC_ID, NEW_PROC),
        set_res(before_list(Group), ([IDLIST|LIST], NEW_PROC)),
        send(test_before(Group, NEW_PROC))),
(test_before(GROUP, []):_
        check_res(proc_group_table, (GROUP, PROCS)),
        check_res(site_group_table, (GROUP, SITE)),
        check_res(before_list(GROUP), (IDlist, _)),
        send(_, _, msg_rqst(SITE, before_list(IDlist, GROUP))))
(test_before(G,P):-print(`Have not received all idlists yet'))
```

Step 7. *Any messages in the wait queue that are also in the before list are transferred to the delivery queue, maintaining their order. The GBCAST message is also placed on the delivery queue.*

The *gbcast_server* requires an additional description of the *before_list* message. Receipt of this by member modules initiates a transfer of the new (*before*) *List* to the respective *delivery_queue* modules (sent as *new_idlist*).

```
(before_list(List, GROUP):-
        check_res(proc_group_table, (GROUP, PROCS)),
        send(send_list(List, PROCS))),
(send_list(List, []):-print(`Updated IDLIST sent')),
(send_list(List, [M|Rest]):-
        send(delivery_queue(M),_, new_idlist(List)),
        send(send_list(List,Rest))),
```

The *delivery_queue* modules perform the necessary comparisons to move entries from the *wait_queue* to the *d_queue*. Finally, a *get_gbcast* message is sent to the respective *priority_queue* modules to obtain a copy of the GBCAST message.

```
(new_idlist(List):-
        check_res(wait_queue(PROC_ID), WAIT),
        check_res(d_queue(PROC_ID), DELIVERY),
        send(compare_queues(WAIT, List,[],WAIT))),
```

```
(compare_queues([], BeforeList, TEMP, WAIT):-
          set_res(wait_queue(PROC_ID), WAIT),
          check_res(d_queue(PROC_ID), DELIVERY),
          set_res(d_queue(PROC_ID), [TEMP\DELIVERY])),
          send(priority_queue(PROC_ID),_,get_gbcast)),
(compare_queues([Top\Rest], BeforeList, N, NewWait)):-
          member(Top, BeforeList),
          append(Top, N, TEMP),
          remove_element(Top, NewWait, WAIT),
          send(compare_queues(Rest, BeforeList, TEMP, WAIT))),
```

The *priority_queues* react to the *get_gbcast* message by removing the GBCAST message from the *waiting_message* resource and transferring it to the *delivery_queue:*

```
get_gbcast:-
        check_res(waiting_message(PROC_ID), (TOP, M, Label,
        gbcast_wait)),
        remove(waiting_message(PROC_ID), (Val, M, Label,
        gbcast_wait)),
        send(_,dq_port(PROC_ID),add_gbcast_message(M, Label)),
```

Step 8. *The contents of the wait queue are appended to the delivery queue, and the wait queue is deleted.*

The *delivery_queue* modules perform cleanup operations after receiving the GBCAST message from the *priority_queue* modules. The message (*MESSAGE*) is received through the *add_gbcast_message* sent in the previous step. The message is added to the delivery queue resource *d_queue* as well as the remaining contents of the *wait_queue,* after which the *wait_queue* is removed. A *clean_gbcast* message is sent to the *priority_queue* modules to remove the GBCAST message from the head of the *p_queues.*

```
add_gbcast_message(MESSAGE):-
        check_res(d_queue(PROC_ID), DELIVERY),
        set_res(d_queue(PROC_ID), [MESSAGE\DELIVERY]),
        check_res(wait_queue(PROC_ID), WAIT)),
        check_res(d_queue(PROC_ID), DELIVERY2),
        set_res(d_queue(PROC_ID), [WAIT\DELIVERY2]),
        remove(wait_queue(PROC_ID))
        send(priority_queue(PROC_ID),_,clean_gbcast)
```

Step 9. *The GBCAST messages are removed from the heads of the ABCAST queues.*

This completes the final step of the protocol. The code shown below is within the *priority_queue* module to clean up the GBCAST message from the *p_queue.* A *check_pq* message is sent as outlined within the ABCAST protocol to commence delivery of any eligible ABCAST messages.

```
clean_gbcast:-
        check_res(p_queue(PROC_ID), [TOP|REST]),
        set_res(p_queue(PROC_ID), REST),
        check_res(p_queue(PROC_ID), [Pri\R]),
        check_res(waiting_message(PROC_ID), (Pri, M, Label, State)),
        send(check_pq(abcast, PROC_ID, Label)))
```

Node model

A node is defined as being composed of a *high_level_services* module (describing the higher-level services), a *generic_protocol_server* module (describing the various supported broadcast protocols), and a *low_level_services* module (describing the low-level support for the node). The *generic_protocol_server* module details were presented in the previous section. This section will present the details for the high- and low-level services support for the modeled node. The DSL structure is shown in Fig. 8.18.

The *high_level_services* module is intended to generate the communication requests (typically from an application running on the node). In this example, a set of process modules residing within this higher-order module is defined to generate various broadcast requests. Each process module possesses a unique *PROC_ID* and a *requests* resource. The latter stores communication requests to be initiated from the process. The resource contains three fields. The first identifies the

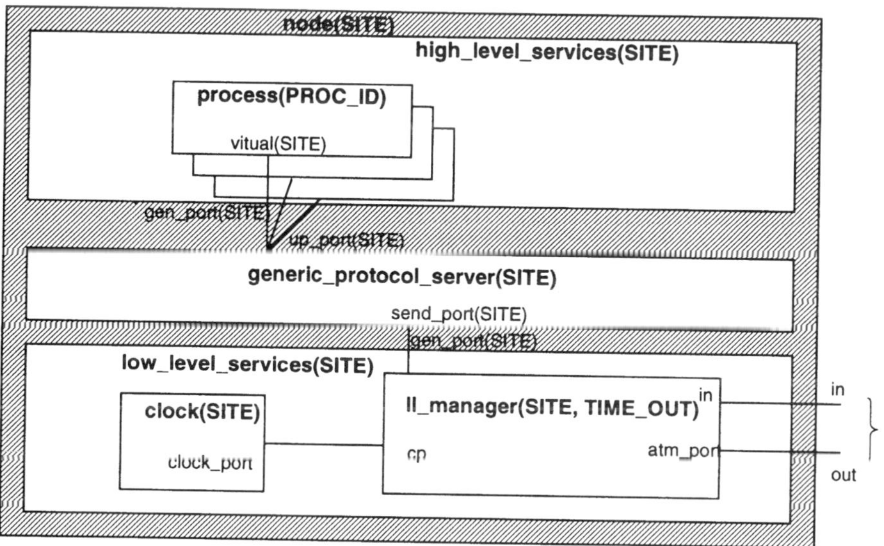

Figure 8.18 Node representation.

protocol requested, the second contains a message to be executed by a destination process, and the third indicates the *PROC_ID* of the destination process. Each process follows a simple behavior of cycling through each of its entries in the *requests* resource and sending a form message down to the *generic_protocol_server.* This behavior is shown below.

```
module(process(PROC_ID), [
(start_process:-
        check_res(requests, (PRO, MSG, DEST)),
        set_res(requests, (PRO, MSG, DEST)),
        delay(PROCESS_DELAY),
        send(_, virtual(SITE), form(PRO, PROC_ID, DEST, MSG))
        send(start_process)]).
```

This description for a process will loop endlessly to provide a sequence of requests. As shown in the broadcast protocol models, the form message will initiate a given broadcast protocol, which will eventually result in an *msg_rqst(DEST_SITE,MESSAGE)* message being sent from the *generic_protocol_server* to the *lower_level_services* module.

The *msg_rqst* message is a low-level message to request a transmission of *MESSAGE* to a node address *DEST_SITE*. The *low_level_services* higher-order module is composed of an *ll_manager (SITE,TIME_OUT)* and a *clock(SITE)* module. The *ll_manager* provides the behavior for actual communication and transmission to an ATM switch, whereas the clock (as in the previous examples) provides the timing requirements for the communication. The *TIME_OUT* parameter identifies a time-out value for the low-level communication protocol. The protocol modeled in this example is a simple send (*hmsg*) and acknowledge (*ack*) protocol. If an acknowledgment is not received within *TIME_OUT* number of ATM cells, then an error message is recorded.

To provide low-level services, the *ll_manager* maintains a *translation_table* resource identifying the ATM (VPI, VCI) addresses for the destination. A message identification is also recorded for each message (from a *message_id* resource) to keep track of outstanding acknowledgments. A *time_out_buffer* for each message is used to record the waiting time for time-out purposes. Finally, a resource (*msg_store*) maintains a copy of messages waiting to be processed. The code for handling *msg_rqst* is given below.

```
msg_rqst([], _).
msg_rqst([DEST_SITE|REST], MESSAGE):-
      check_res(translation_table(DEST_SITE), (VPI,VCI)),
      check_res(message_id, VALUE),
      COUNT is VALUE+1,
      set_res(message_id, COUNT),
```

```
        create(msg_store(COUNT), (VPI, VCI, DEST_SITE, MESSAGE)),
        send(msg_rqst(REST, MESSAGE)
```

As defined in the ATM model, a *new_cell* message is sent to and from the switch. Hence the *ll_manager* must process such a message. This is defined by

```
new_cell(VPI, VCI, MESSAGE):-
        send(MESSAGE).
```

The convention in this example is that *MESSAGE* can be in the form of either a new message sent from another node—described as *hmsg(Sender, Destination, Message_ID, Message)*—or an acknowledgment of a past message (*ack(Sender, Destination, Message_ID)*). In the case of an *hmsg* message, an acknowledgment is sent back to the sender and the contents are forwarded to the *generic_protocol_server.* In the case of an *ack* message, the respective message is removed from the *time_out_buffer,* indicating a successful message transfer. The DSL descriptions are given below:

```
hmsg(Sender, _, MSGID, MESSAGE):-
        check_res(translation_table(Sender), (VPI, VCI)),
        send(_, atm_port(SITE), new_cell(VPI, VCI, ack(SITE, Sender,
        MSGID))),
        send(_, gen_port, MESSAGE)
ack(SITE, Sender, MSGID):-
        remove(time_out_buffer(MSGID), (TIME, DATA)),
        print(Message successfully received at destination').
```

There is also a *clock* message sent from the *clock* module to initiate the transmission of a cell. If a message is awaiting transmission, it is sent as an *hmsg* message to the destination; otherwise a *null* ATM cell is sent to the switch. The *clock* message is also used to update any time-out values for waiting responses (this is performed within the *update_times* behavior shown below). The description of the *clock* module is the same as the one given for the ATM switch design and will not be repeated here.

```
clock:-
        check_res(msg_store(MSGID), (VPI, VCI, DEST_SITE, MESSAGE)),
        remove(msg_store(MSGID), DATA),
        create(time_out_buffer(MSGID), (TIME_OUT, DATA)),
        send(_,atm_port(SITE), new_cell(VPI, VCI, hmsg(SITE,
        DEST_SITE, COUNT, MESSAGE)),
        send(update_times)
clock:-
        send(_, atm_port(SITE), new_cell(null,null,null),
        send(update_times)
update_times:-
        check_res(time_out_buffer(MSGID), (TIME, DATA)),
        NEW is TIME-1,
```

```
        set_res(time_out_buffer(MSGID), (NEW, DATA)), fail.
update_times:-.
        check_res(time_out_buffer(MSGID), (TIME, DATA)),
        TIME<0,
        print(`Message`),print(MSGID),print(`timed out'), fail.
update_times:- print(`Timeouts updated').
```

ATM—Broadcast System Model

The DSL models presented in the ATM and this section provide the basic building blocks for defining an ATM-based broadband network running the defined broadcast protocols. This section will demonstrate a design validation and simulation exercise that is typical for many protocol designs. A network model depicting a configuration based upon the network introduced in the ATM section and the nodes defined in this section are shown in Fig. 8.19. DSL model details for this level of the example will not be dwelt upon. The network is assumed to provide a bidirectional virtual path connection among the three nodes.

Each node is assumed to contain ten processes running within its *high_level_services* module. The processes are identified (*PROC_ID*) by a combination of the node number (*SITE*) and a unique integer *n* greater than or equal to 1, but less than 10. The resulting *PROC_ID* is in the form *SITE.n*. Five process groups (G1...G5) are also defined within the model, and each process has 20 predefined broadcast messages consisting of a mix of the three broadcast protocols. Each process randomly chooses one of these messages, at random intervals,

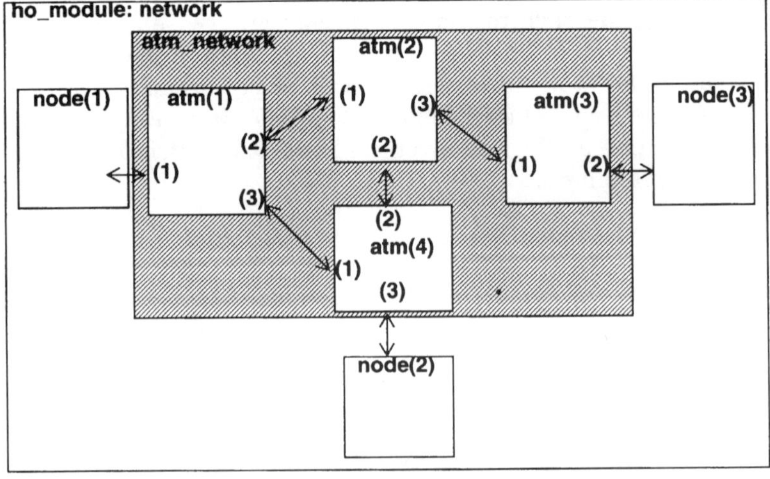

Figure 8.19 Combined network model.

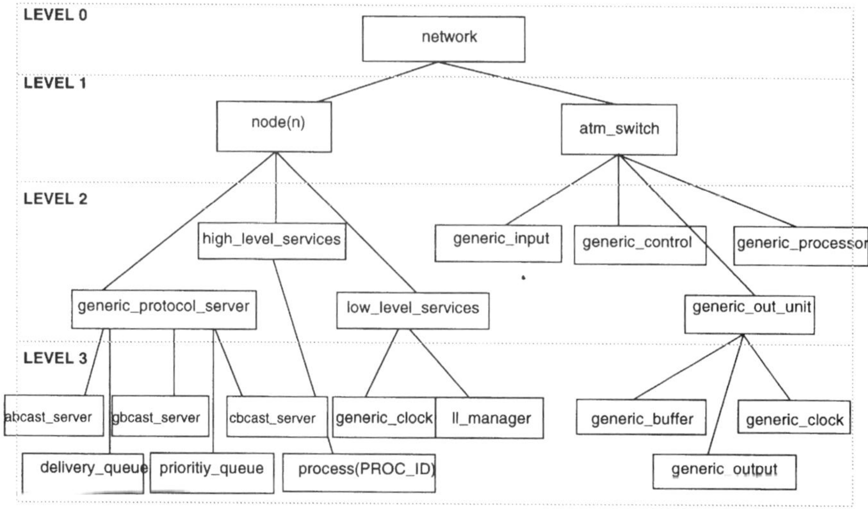

Figure 8.20 Hierarchy tree for combined model.

to broadcast to the various destinations. A constraint is also defined for the queue resource in the *generic_buffer* modules within each ATM switch. The constraint defines a maximum output queue size to detect overflowing message buffers (which is one of the major design issues in ATM switches).

When the model is loaded within DASE, a hierarchy tree is generated by the system, as shown in Fig. 8.20. The protocol modules are represented at the lowest level of the module hierarchy (level 3). Hence it may be desirable to implement these modules in software. All nodes are initiated with messages which randomly generate broadcasts; hence each simulation run can provide different outputs. A typical output is shown in Fig. 8.21. The effects of different observational levels can be seen. There are concurrent broadcast messages generated by different processes, though the output shows only those from process (1.1) residing on site 1. The ABCAST message is sent to processes at sites 2 and 3.

The first segment of code is set at an observation level of 1, which restricts the view to the higher-level protocols being modeled. The sample code shows an ABCAST message request from the higher-level protocol (11), the receipt of the message at one of the destinations (12), update of the priority queue (13), and a suggestion of a priority value back to the destination (14). This view avoids most of the low-level details of the ATM protocol. However, it provides a reasonable view of the message generation and delivery for the different broadcast protocols.

```
>set_level(1).

observation level set to 1

>run.

DSL SIMULATOR: Scheduling message from hls(1,200) via gps(1) to abcast(1)
Source Port: virtual(1,200) Message is: form(abcast,200,[3,2],abm3)

< 0.0 > DSL SIMULATOR: new entry added to message queue of abcast(1)
< 0.0 > DSL PROCESSOR: Message start_process completed at module hls(1,200)
< 0.0 > DSL SIMULATOR: Module abcast(1) busy (delayed) until: 1.0
< 1.0 > DSL SIMULATOR: abcast(1) resource updated: label_count to value: 2
 Sending an abcast message abm3

DSL SIMULATOR: Scheduling message from abcast(1) via gps(1) to lls(1,20)
Source Port: lport(1) Message is: msg_rqst([3,2],abcast(abm3,2,200,1))

DSL SIMULATOR: Scheduling message from lls(2,20) via gps(2) to abcast(2)
Source Port: gen_port Message is: abcast(abm3,2,200,1)

< 10.0 > DSL SIMULATOR: new entry added to message queue of abcast(2)
< 10.0 > DSL PROCESSOR: Message hmsg(1,2,2,abcast(abm3,2,200,1)) completed at module lls(2,20)

DSL SIMULATOR: Scheduling message from abcast(2) via abcast(2) to priority_q(2)
Source Port: pq_port(2) Message is: add_new(abcast,1,abm3,2,200)

< 10.0 > DSL SIMULATOR: new entry added to message queue of priority_q(2)
< 10.0 > DSL PROCESSOR: Message abcast(abm3,2,200,1) completed at module abcast(2)
< 10.0 > DSL SIMULATOR: Module priority_q(2) busy (delayed) until: 11.0
< 10.0 > DSL SIMULATOR: input(2,4) resource updated: cell_buffer to value: 1 , 2 , ack(2,1,2) , unprocessed

DSL SIMULATOR: Scheduling message from input(2,4) via input(2,4) to control(4)
Source Port: control_port(4,2) Message is: new_cell(2)

< 10.0 > DSL SIMULATOR: new entry added to message queue of control(4)
< 10.0 > DSL PROCESSOR: Message new_cell(1,2,ack(2,1,2)) completed at module input(2,4)

DSL SIMULATOR: Scheduling message from lls(3,20) via gps(3) to abcast(3)
Source Port: gen_port Message is: abcast(abm3,2,200,1)

< 10.0 > DSL SIMULATOR: new entry added to message queue of abcast(3)
< 10.0 > DSL PROCESSOR: Message hmsg(1,3,1,abcast(abm3,2,200,1)) completed at module lls(3,20)

DSL SIMULATOR: Scheduling message from abcast(3) via abcast(3) to priority_q(3)
Source Port: pq_port(3) Message is: add_new(abcast,1,abm3,2,200)

DSL SIMULATOR: Scheduling message from priority_q(3) via priority_q(3) to abcast(3)
Source Port: abcast Message is: updated_priority(abcast,1,2,_1344)

< 11.0 > DSL SIMULATOR: new entry added to message queue of abcast(3)
< 11.0 > DSL PROCESSOR: Message add_new(abcast,1,abm3,2,200) completed at module priority_q(3)

DSL SIMULATOR: Scheduling message from priority_q(2) via priority_q(2) to abcast(2)
Source Port: abcast Message is: updated_priority(abcast,1,2,_7436)

< 11.0 > DSL SIMULATOR: new entry added to message queue of abcast(2)
< 11.0 > DSL PROCESSOR: Message add_new(abcast,1,abm3,2,200) completed at module priority_q(2)

broadcast send completed for message msg_rqst(1,abcast(1,sug_priority(_11416),2))
< 11.0 > DSL PROCESSOR: Message updated_priority(abcast,1,2,_11416) completed at module abcast(2)

yes
```

Figure 8.21 Simulation output.

```
>set_level(4).

observation level set to 4

>run.

DSL SIMULATOR: Scheduling message from lls(1,20) via site(1) to input(1,1)
Source Port: atm_port(1) Message is: new_cell(2,1,hmsg(1,2,2,abcast(abm3,2,200,1)))

< 2.0 >  DSL SIMULATOR: new entry added to message queue of input(1,1)

DSL SIMULATOR: Scheduling INTERNAL message from lls(1,20)
Message is: update_times

< 2.0 >  DSL PROCESSOR: Message clock completed at module lls(1,20)
< 2.0 >  DSL SIMULATOR: input(1,1) resource updated: cell_buffer to value: 2 , 1 , hmsg(1,2,2,abcast(abm3,2,200,1)) , unprocessed

_24796  control_port(1,1)  new_cell(1)
 Source Destin, and Sender input(1,1)control(1)input(1,1)
Module is control(1) message is new_cell(1)
Checking method in control(1)

Checking method in generic_control(1)
DSL SIMULATOR: Scheduling message from input(1,1) via input(1,1) to control(1)
Source Port: control_port(1,1) Message is: new_cell(1)

< 2.0 >  DSL SIMULATOR: new entry added to message queue of control(1)
< 2.0 >  DSL PROCESSOR: Message new_cell(2,1,hmsg(1,2,2,abcast(abm3,2,200,1))) completed at modulo input(1,1)

DSL SIMULATOR: Scheduling message from control(1) via control(1) to input(1,1)
Source Port: input_port(1,1) Message is: read_rqst

< 2.0 >  DSL SIMULATOR: new entry added to message queue of input(1,1)
< 2.0 >  DSL PROCESSOR: Message new_cell(1) completed at module control(1)
< 2.0 >  DSL SIMULATOR: input(1,1) resource updated: cell_buffer to value: 2 , 1 , hmsg(1,2,2,abcast(abm3,2,200,1)) , processed

DSL SIMULATOR: Scheduling message from input(1,1) via input(1,1) to control(1)
Source Port: control_port(1,1) Message is: cell_info(2,1,hmsg(1,2,2,abcast(abm3,2,200,1)))

< 2.0 >  DSL SIMULATOR: new entry added to message queue of control(1)
< 2.0 >  DSL PROCESSOR: Message read_rqst completed at module input(1,1)
control unit: transferring a cell to output

DSL SIMULATOR: Scheduling message from control(1) via control(1) to processor(1)
Source Port: proc_port(1) Message is: new_message(2,1,hmsg(1,2,2,abcast(abm3,2,200,1)))

< 2.0 >  DSL SIMULATOR: new entry added to message queue of processor(1)
< 2.0 >  DSL PROCESSOR: Message cell_info(2,1,hmsg(1,2,2,abcast(abm3,2,200,1))) completed at module control(1)
ATM processor: routing a cell

DSL SIMULATOR: Scheduling message from processor(1) via processor(1) to buffer(1,3)
Source Port: out_port(3) Message is: atm_cell(2,3,hmsg(1,2,2,abcast(abm3,2,200,1)))

< 2.0 >  DSL SIMULATOR: new entry added to message queue of buffer(1,3)
< 2.0 >  DSL PROCESSOR: Message new_message(2,1,hmsg(1,2,2,abcast(abm3,2,200,1))) completed at module processor(1)
< 2.0 >  DSL SIMULATOR: buffer(1,3) resource updated: queue to value: [(2 , 3 , hmsg(1,2,2,abcast(abm3,2,200,1))),(null , null , null)]
< 2.0 >  DSL PROCESSOR: Message atm_cell(2,3,hmsg(1,2,2,abcast(abm3,2,200,1))) completed at module buffer(1,3)

DSL SIMULATOR: Scheduling message from buffer(1,3) via buffer(1,3) to output(1,3)
Source Port: output_port(1,3) Message is: next_info((2 , 3 , hmsg(1,2,2,abcast(abm3,2,200,1))))

DSL SIMULATOR: Scheduling message from output(1,3) via switch(3) to input(1,3)
Source Port: out_port(1,3) Message is: new_cell(2,3,hmsg(1,2,2,abcast(abm3,2,200,1)))
```

Figure 8.21 *(Continued)*

The second segment of the code shows the same simulation run with a different observation level. In this scenario, the switching function is of interest; hence higher-level protocol messages are not observed. The code segment shows the receipt of a cell from site 1 to a switch's port (15), translation of the cell headers (16), and routing to the destination port on the switch (17).

As the examples try to illustrate, different observation levels during simulation can help the user focus on the desired level of detail. This in turn, permits the user to zoom in on different segments of the model during the course of the simulation, permitting observation of pertinent information. The real impact of these features is difficult to depict in words—the trial and use of the tool is the best way of observing the benefits.

Chapter

9

Advanced Issues

Architectural-level modeling and design of complex systems is ripe for future research. This chapter introduces some of the outstanding issues that have not been covered in the book. These will enhance an environment's ability to aid the designer.

Model Base

The key to achieving good interaction between the environment and the designer is the use of a reasonable model base. The environment must be able to access model information in a fast and efficient manner. In order to facilitate this, the model library needs to be organized in an intuitive manner. The domain-directed approach ensures that information can be grouped into meaningful classes. While there are no clear-cut rules to guarantee that the information is appropriate for any design organization, some points of consideration will help an organization decide on a suitable strategy.

Classification

The first consideration is to classify components of the design domain. The classification is an outgrowth of relationships between models. A relationship between models has already been introduced in the form of a composite model called the higher-order module. By definition, such a model implies an intricate association with its constituent models. We have been introduced to a library system capable of organizing modules in terms of their relationships within higher-order modules. This is necessary for model construction and simulation.

Another concern is related to aiding the designer in exploration of the solution space. If at some point in time during simulation of the DSL model a severe design violation is detected, the environment should be able to interject and suggest some alternative solutions. To perform such a function, the environment also needs to have access to additional insight. The knowledge needed to perform this is related to the ability to identify two or more models as substitutes. Such models are said to possess a *model equivalency.* Without diverging into the world of theoretical computer science, we shall informally define model equivalency as follows:

> DEFINITION: Two or more modules are said to be *model equivalent* if their behavior in response to a set of messages is indistinguishable by an external module.

For purposes of this discussion, model equivalency is a black-box view of modules in which only the events occurring at the inputs and outputs of the modules are observable. Once models are identified as being equivalent, they can be maintained in a class that can be easily accessed by the environment. For example, a set of microprocessors may be defined as being composed of different brands of processors, all possessing the same functionality and therefore grouped into the same class. Alternatively, a class describing different sorting algorithms could be created.

Determining whether two models are equivalent is tricky. Figure 9.1 shows two modules. Each module has a set of possible messages which it can process. These are indicated by the sets *A* and *B*. The union of the two sets implies a common set of messages which both modules are capable of interpreting. This subset (*C*) consists of common messages that can be observed by another module and therefore constitutes one condition for determining equivalency.

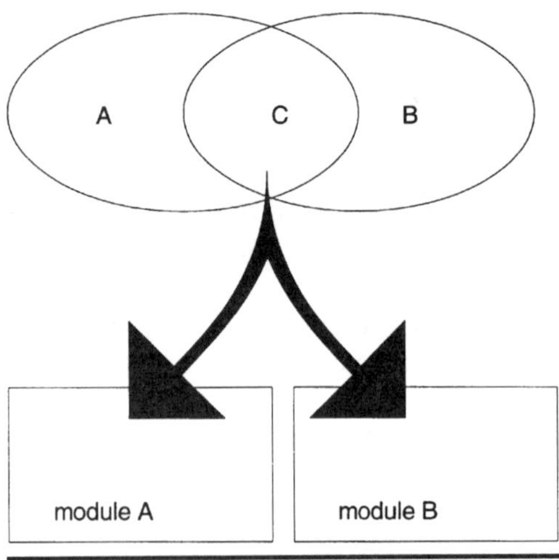

Figure 9.1 Message sets for modules.

Determining a common set of messages is not enough. The output generated by each message belonging to the set of messages C must be the same for both modules. A possible way in which this can be tested is to examine the reachability trees of each module's Petri-net representation. If a path exists from each input place to the same output places, then this may imply similar behavior. However, the outcome must be the same independent of the sequence of messages—i.e., if an input message m_{in} generates an output message m_{output}, this must always be the outcome, regardless of any combination of other input messages.

Checks for equivalence can be performed by the environment prior to grouping models into classes. Additional domain-specific heuristics will be required to aid the environment in organizing the model library. The heuristics may be required to differentiate characteristics of the devices that are being modeled. For example, it may be important to group models of devices that are within a specific cost range, even though their behavior can be termed identical (for example, a faster microprocessor can replace a slower one at a higher cost). Hence a grouping based first upon cost, then on behavior can be established. This helps reduce access times in large model libraries.

Visualization

Data visualization is a fertile research area which can greatly affect design environments. Cues which can signal some important event in a model during execution can be beneficial in understanding the functionality of the system. This in turn helps validate the design and communicate its behavior to other people. Some of the most popular modes of visualization are as follows:

- *Text* is viewed as a primitive form of visualization; however, it forms the basis of any environment. Some elements are best seen through plain text. In addition, no matter how much graphical support an environment may possess, the user must eventually code some aspect of the behavior.
- *Graphics* are a powerful aid in constructing and viewing models. The drawback of these approaches is that as designs become more complex, the usefulness of a graphical representation of the design diminishes. A hierarchical view of the system helps reduce the clutter of the model.
- *Animation* has been a popular feature of simulation programs in recent years. Traditionally thought of as a feature to communicate the model to the management, it is now also perceived as helpful in visualizing some of the major system activities.

- *Virtual reality systems* also promise some relief in visualizing the complexity. The benefits of such systems could be more interactive design exploration with the environment and powerful communication and training tools.

Design Solution Space

The environments discussed have aided in finding models capable of operating within design specifications. These solutions are not necessarily optimal ones. To address this, further research into algorithms which can give metrics on how good a particular design solution is will be necessary. However, there are also some aspects of exploring the design space that have not been discussed that will be mentioned below.

Design space exploration is a decision-making process. It will implicate the designer, the environment, or both. Searching for a solution for a given design specification is analogous to the backtracking mechanism Prolog utilizes. A similar approach (which can be specified in Prolog) can be used to guide the user through an environment. For example, Fig. 9.2 depicts a design exploration path taken for a hypothetical design execution activity. The nodes represent a model of the system. A move from one node to the next represents a modification

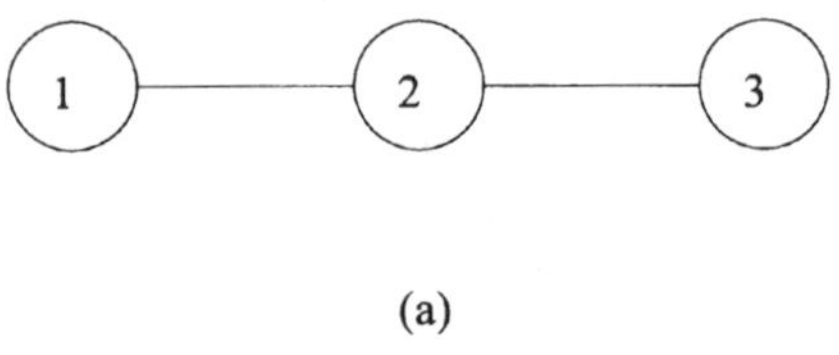

(a)

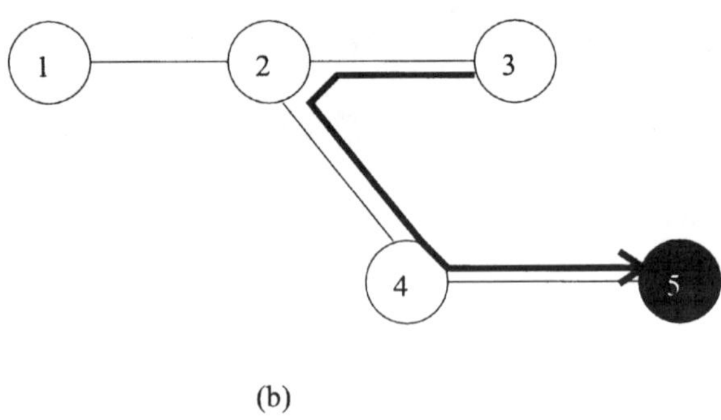

(b)

Figure 9.2 Design backtracking example.

or refinement of the previous node. Each node is subjected to simulation, based upon which it is refined or a solution is found. In the first example, a design has progressed from node 1 to node 2. Node 2 required a modification, resulting in a model represented by node 3. After simulation, node 3 was found not to satisfy the design specifications; however, neither the user nor the environment could offer a substitution for the nonconforming component(s). As a result, the design is backtracked to the model configuration of node 2. At this point (shown in Fig. 9.2*b*), an alternative solution is found, resulting in node 5. This reasoning can continue until a solution is found or no other possible choices are available to the environment.

Following this reasoning, representation of the design exploration algorithm is natural in Prolog. The system specification and constraints can be viewed as the goal of the Prolog program. The representations of the nodes are facts. Design exploration is then reduced to satisfying the goal through the right mix of facts. Additional facts are added by the user or the environment as the activity progresses, resulting in an increase in the overall facts available.

Experimental Results

This section summarizes the results of experiments performed using the DASE tool. The experiments are presented to elaborate on some of the simulation performance capabilities of the DSL simulator to give an appreciation of the length of time needed to simulate designs. The first simulation is composed of several simulations of the DTSS switch example presented in the earlier chapters. The second simulation is that of the ATM switch case study presented in Chap. 8. All simulations were performed on a Sun 10/30 platform with 64 Mbytes of RAM. The workstation was dedicated to the simulations. The resulting figures include operating system overhead.

DTSS simulation

A DTSS switch was simulated with different numbers of subscribers. The numbers of subscribers were 2, 64, 640, 64,000, and 126,000. Telephone traffic was simulated to last for exponential intervals between 2 and 8 seconds per call. Each simulation run was for a simulated period of 15 minutes of switching activity.

Figure 9.3 presents the performance results for the example. The time to simulate increases linearly as the complexity (number of telephone subscribers) increases. At the extremes, a two-subscriber interaction can be simulated in under 13 minutes of CPU time, whereas a 126,000-subscriber simulation will take more than 33 hours of simulation time.

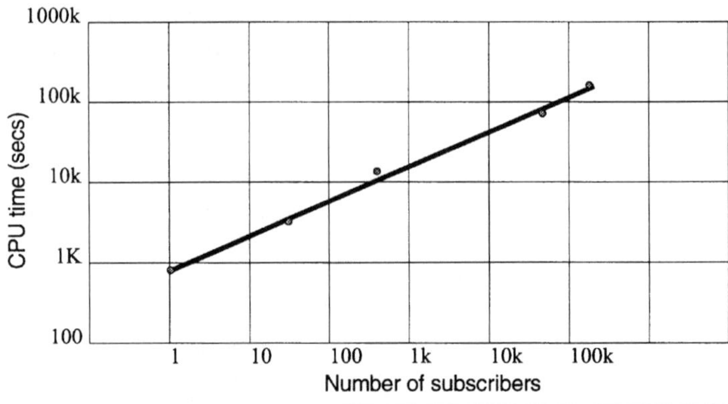

Figure 9.3 Fifteen-minute simulation of voice traffic.

ATM simulation

The model of the ATM switch described in Chap. 8 was simulated. The simulation consisted of different switch sizes of 2, 4, 8, and 16 input/output ports. Each port was loaded by a continuous traffic stream of ATM cells at a rate of 155 Mbits/s. This results in an effective cell rate of 2.8 μs/cell. A recording of the CPU time for each configuration was taken at 2.8 and 8.4 s of simulated time. Approximately 1 million cells were generated at each port every 2.8 s.

The results are plotted in Fig. 9.4. As can be observed, the simulation times are generally extensive. However, the simulations indicate

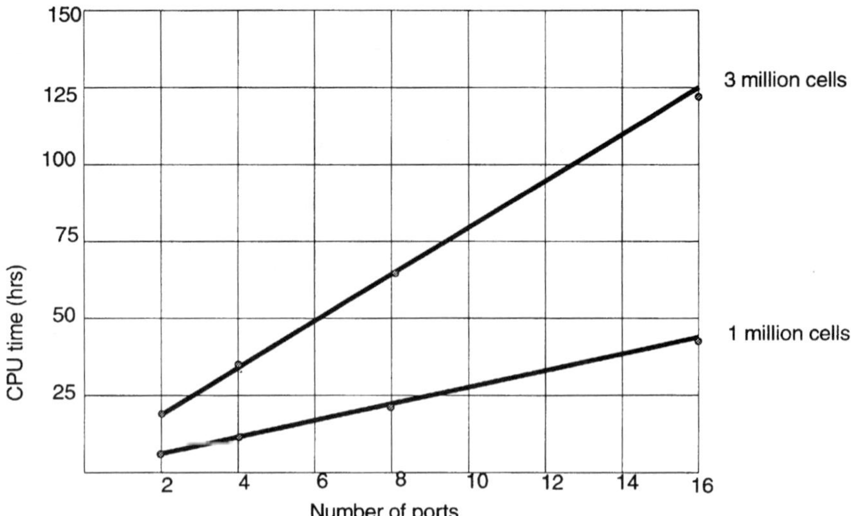

Figure 9.4 Simulation of different ATM switch sizes.

a linear progression of times as model complexity increases. The simulator performs reasonably well; however, for demanding simulations such as the ATM switch, accelerated simulation techniques would be desirable in order to obtain quicker results. Designer experience and knowledge are truly required in order to reduce unnecessary simulation runs and focus on the important aspects of a design.

Appendix

DSL Code Listing

/***

This is the DSL interpreter Version 1.0 for the Design Specification Language written by Oryal Tanir at McGill University. It incorporates the DSL processor and simulator functionalities with synthesis hooks. Special mention is due to Bell Canada for sponsoring a significant portion of these activities. The two major components are the simulation engine and the DSL processor. The simulator allows the scheduling of a timed execution of messages, while the processor allows the construction and manipulation of DSL entities. A significant portion of the code is actually the supporting predicates for implementation of the *send* predicate. The predicate actually performs most of the message-transferring functions. It utilizes several other predicates to handle the proper delivery of messages.

mod_type determines the type of module and its name (if not given) from port connections.

ioa_hierarchy determines the lowest-level destination module and message.

module returns the behavior associated with a given module.

get_method extracts the behavior associated with the message. This is a linear search through the module's list of behaviors.

The *send* predicate has several forms; hence the need for explicit resolution of the sender–destination–port combinations. The complexity of the backtracking is reduced by the list operator in the "dest" parameter. The only other special case that is resolved is the '*' broadcast operator.

For syntactical elegance, we require broadcast specified as "[*]". The resolution of this case is nontrivial. All destination modules that have a path from sender are resolved. This is contained within the *create_dest_list* predicate. However, the list is prepruned using the *get_method* and *isa_hierarchy* predicates to evaluate whether the resolved modules can indeed understand the message. If a module cannot, it is removed from the list.

The pruned list of possible modules is returned as a list to *queue_list_of,* which consequently extracts the elements of the list and queues them for processing by the modules. */

```
send(-, S, M):-sendm(X, S,M).
send([*], S, M):-
                    current_module_processed(Source),
                    create_dest_list(Source, S, M, []).
create_dest_list(Source, S_Port, Message, LIST):-
                    path(Source, DEST, [S_Port, NOCARE]),
                    int_mod_type(Source, DEST, S_port, Sender,Rec,
                    SPort),
                    mod_type(Sender, Rec, SPort, Module),
                    not(member(Module, LIST)),
                    isa_hierarchy(Module, Module1),
                    module(Module1, MethodList),
                    get_method(Message, MethodList, Method),
                    create_dest_list(Source, S_Port, Message,
                    [Module|LIST]).
create_dest_list(S, P, M, List):-
                    queue_list_of(List, M).
queue_list_of([Module|List], M):-
                    print(´ Broadcast message sent to ´),
                    print1(Module),
                    queue_message(Module, M),
                    queue_list_of(List, M).
queue_list_of([], M):-
                    print(´ broadcast send completed for
                    message ´),print1(M)
```

/**

The following is the second case for the *send* predicate. In this form, we anticipate a list of destinations from "empty" ([]) to valid destinations. The latter require a connection to be "live" between the sender and the destination (this implies a *path* statement between the two to be defined).

**/

```
send(DEST, S, M):-
                    list_length(DEST, Len),
                    Len>1,
                    top_of_list(DEST, TOP, REST),
                    sendm(TOP, S, M),
                    send(REST, S, M).
send(DEST, S, M):-
                    list_length(DEST, Len),
```

```
                        LEN = 1,
                        top_of_list(DEST, TOP, REST),
                        sendm(TOP, S, M).
send([], S, M):-
                        sendm(X, S, M).
top_of_list([H|T], H, T).
send(D, S, M):-sendm(D, S, M).
```

/**

Finally, we have the plain-vanilla send case *sendm*. At this point, the higher-level *send* predicates have already preprocessed the destination list and provided a unique destination.

**/

```
sendm(Destin,S_port,Message):-current_module_processed(Source),
                        print('\n- = = ——————————————————————————\n'),
                        int_mod_type(Source, Destin, S_port,
                        Sender,Rec, SPort),
                        mod_type(Sender, Rec, SPort, Module),
                        isa_hierarchy(Module, Module1),
                        module(Module1), MethodList),
                        get_method(Message, MethodList, Method),
                        print('DSL SIMULATOR: Scheduling message from
                        '),
                        print(Source), print(' via '), print(Sender),
                        print(' to '),
                        print(Module),
                        print('Source Port: '),
                        print(S_port),
                        print(' Message is: '), print(Message),
                        print('- = = ——————————————————————————\n'),
                        queue_message(Module, Message).
send(Message):-
                        print('\n——————————————————————————-\n'),
                        current_module_processed(Module),
                        print('DSL SIMULATOR: Scheduling INTERNAL
                        message from '),
                        printl(Module),
                        print('Message is: '), printl(Message),
                        print('——————————————————————————\n'),
                        queue_message(Module, Message).
```

/* The following are pretty printing routines to format the output ***/

```
pwrite(X):-deb(on), write(X).
pwrite(X).
deb(off).
don:- retract(deb(X)), asserta(deb(on)).
doff:-retract(deb(X)), asserta(deb(off)).
```

/**

Determine if the destination is a module or a higher-order module: This predicate is required to determine if a given destination module is a higher-order module. If it is, a "final" destination is resolved that is a module. This is necessitated by the fact that higher-order mod-

ules provide structural information; hence any behavior to be executed at the destination can be derived only from the behavior list of a module.
**/

```
int_mod_type(Source,Destin, Port, Source, Destin, Port):-
                        path(Source,Destin,[Port,_]),
                        module(Destin,_).
int_mod_type(Source,Destin, S_port, Parent, Target, TP):-
                        path(Source, Destin, [Port, D_port]),
                        ho_module(Destin, List),
                        member(Source, List),
                        path(Source, Destin, [Port, D_port]),
                        path(Destin, Next, [D_port, N_port]),
                        not(equalt(Source,Next)),
                        not(member(Next, List)),
                        int_mod_type(Destin, Next, D_port, Parent,
                        Target, DPort).
int_mod_type(Source,Destin, S_port, Destin, Next, D_port):-
                        path(Source, Destin, [Port, D_port]),
                        ho_module(Destin,_),
                        path(Destin, Next, [D_port, N_port]),
                        int_mod_type(Destin, Next, D_port, Dest, New,
                        Port_D).
equalt(X,X).
mod_type(Source,Dest, Port, Dest):-module(Dest,_).
mod_type(Source,Dest, Port, Module):-ho_module(Dest,_),
     path(Source, Dest,[Port,Dest_port]),
     path(Dest,Module,[Dest_port,_]),
                        mod_type(Dest, Module, Dest_port, Module).
```

/**

Search the ancestors of the module to find the method:

**/

```
isa_hierarchy(Module,Module).
isa_hierarchy(Module,Module1):-
     isa(Module, Module2),
     isa_hierarchy(Module2, Module1).
```

/**

get_method finds the appropriate method to execute.
A behavior list is searched until a header that matches a given message name is found.

**/

```
get_method(Message,[FirstMethod|Rest], Method):-
     decode(Message, FirstMethod, Method).
get_method(Message, [_|Rest], Method):-
     get_method(Message, Rest, Method).
```

/**

decode decodes the message to be executed (a part of *get_method*).

**/

```
decode(Message, Message, true).
decode(Message,(Message:-Body), Body).
```

/***

queue_message stores the message in a module's internal queue.

***/

```
queue_message(Mod1, Behavior):-
                        sim_time(Time),
                        retract(message_queue(Mod1,Elements)),
                        app(Elements, [(Behavior,Time)], New_q),
                        asserta(message_queue(Mod1, New_q)),
                        update_next(Mod1).
```

/***

APPEND ROUTINE

This routine is added for ease of portability between different Prolog environments. It mimicks the *append* predicate available on some Prolog systems.

***/

```
app([],X,X).
app([X1|Z],Y,[X1|X2]):-app(X2,Y,Z).
```

/***

AUXILIARY PREDICATES

print is a hierarchical information output mechanism in DSL. The *evaluate_level* predicate checks to see if the current module processed is actually higher up than the desired observation level of the simulator tree. If so (defined by *dsl_level*), the *write* output is observed. This selectively allows the observation of different levels of simulation.

***/

```
dsl_level(-1).
set_to(L):-set_level(L), retract(mark(F)), asserta(mark(on)).
reset:-retract(mark(F)), asserta(mark(off)).
mark(off).
print1(X):-current_module_processed(Module),
                    model_level(Module, Mlevel), mark(FLAG),
                    evaluate_level2(Mlevel, X, FLAG),!.
print(X):-current_module_processed(Module),
                    model_level(Module, Mlevel),mark(FLAG),
                    evaluate_level(Mlevel, X, FLAG),!.
evaluate_level(MLevel, X, off):-
                    dsl_level(LEVEL),
                    MLevel =<LEVEL,
                    write(X).
evaluate_level(MLevel, X, on):-
                    dsl_level(LEVEL),
                    MLevel = LEVEL,
                    write(X).
evaluate_level(MLevel, X, FLAG).
evaluate_level2(MLevel, X, on):-
```

```
                        dsl_level(LEVEL),
                        MLevel = LEVEL,
                        write(X),nl.
evaluate_level2(MLevel, X, off):-
                        dsl_level(LEVEL),
                        MLevel = <LEVEL,
                        write(X),nl.
evaluate_level2(MLevel, X,F).
```

/**
If the *next_time* of a module is -1 (i.e., no messages are scheduled), then it gets updated to the current simulation time.
**/

```
update_next(Modulex):-
next_time(Modulex, Temp),
                  Temp = = -1,
                  sim_time(Time),
                  retract(next_time(Modulex,_)),
                  asserta(next_time(Modulex, Time)),print('\n'),
                  print_sim_time,
                  print('DSL SIMULATOR: new entry added to
                  message queue of'),
                  print(Modulex).
                  update_next(Modulex):-print('\n').
```

/*******Delay predicate:***
Delays related to modules are represented and logged by the delay predicate. The predicate detects a *delay* parameter and suspends operations on that module until the delay time (delta) has elapsed.

The flow control is managed by the use of *module_state(X)* predicates. When X = *ready,* the module is in a normal mode of operation (i.e., its operation has not been suspended). If X = *delayed,* the module is sleeping, and will awake when the delayed time has elapsed.
**/

```
delay(Delta):- current_module_processed(Module),
                    module_state(Module, delayed),!,
                    retract(module_state(Module,_)),
                    asserta(module_state(Module, ready)).
delay(Delta):- current_module_processed(Module),
                    sim_time(T1),
                    retract(module_state(Module,_)),
                    asserta(module_state(Module,delayed)),
                    retract(next_time(Module,_)),
                    NN is (T1 + Delta),
                    print_sim_time,
                    print('DSL SIMULATOR: Module'),print(Module),
                    print(' busy (delayed) until: '),
                    printl(NN),
                    asserta(next_time(Module,NN)),!, fail.
```

/**
The following is a pretty-print format for printing out the simulation time at the leftmost position on the output.
**/

```
print_sim_time:- sim_time(T), print('<'),print(T),print('> ').
```

/********SET PREDICATE***************************************
This predicate sets the variable *X* to *Y* of resource *R*—that is, *set_res(R,Y)*. Resources are specified internally by DSL in the *(R,Y)* notation, where *R* is the resource name. There is an association created internal to DSL that links the resource name to a module name (the module that invoked the *set* command).
**/

```
set_res(Resource_Name, Value):- current_module_processed(Module),
                    retract(resource(Module, Resource_Name,
                    Old_value)),
                    asserta(resource(Module, Resource_Name,
                    Value)),
                    print_sim_time
                    print('DSL SIMULATOR: '),print(Module),print
                    (' resource updated: '),
                    print(Resource_Name),
                    print(' to value: '),printl(Value).
```

/**********CHECK resource***********************************
This predicate returns "Value" assigned to "resource."
**/

```
check_res(Resource, Value):- current_module_processed(Module),
                    resource(Module, Resource, Value).
```

/**********Resource creation and removal predicates*********/

```
remove(R, V):- current_module_processed(Module),
      retract(resource(Module, R, V)).
create(R, V):- current_module_processed(Module),
      asserta(resource(Module, R, V)).
```

/**********Probe predicate**********************************
This returns the value *DATA* available on a persistent port *PORT*.
**/

```
probe(PORT, DATA):- current_module_processed(Module),
                    path(Module, ModuleX, [PORT, PortX]),
```

```
                    resource(ModuleX, persistent_port, (PortX,
                    DATA)).
```

/***

List-processing predicates are defined below. These are intended to provide sufficient functionality to define data structures such as queues and stacks.

list-length(LIST,LENGTH) expects a list and evaluates the number of elements in it.

remove_element(LIST, key, RESULT) finds *key* in *LIST* and removes it, resulting in *RESULT*.

last_list(LIST,LAST) returns the last element in the list.

iqsort(LIST, SORTED) is the improved Quicksort algorithm of Clocksin and Mellish (1984).

***/

```
list_length([],0).
list_length([Head|Tail], K):- list_length(Tail, J), K is J+1.
remove_element([Key|Tail], Key, Tail).
remove_element([H|T], Key, [H|R]):- remove_element(T, Key, R).
last_list([Head|[]], Head).
last_list([Head|Tail, H):- last_list(Tail, H).
partition([XT|Tail], Pivot, [X|Before], After):-
                    X=<Pivot,
                    !,
                    partition(Tail, Pivot, Before, After).
partition([X|Tail], Pivot, Before, [X|After]):-
                    x>Pivot,
                    !,
                    partition(Tail, Pivot, Before, After).
partition([],_,[],[]).
iqsort(List, Result):-
                    iqsort_aux(List, [], Result).
iqsort_aux([X|Tail], Stack, Result):-
                    !,
                    partition(Tail, X, Before, After),
                    iqsort_aux(After, Stack, NewStack),
                    iqsort_aux(Before, [X|NewStack], Result).
iqsort_aux([], Stack, Stack).
/*********Library handling routines************************/

load_dsl:- system('clear'),system('date'),nl,
                    set_up, load_dsl1,
                    system('clear'),system('date'),nl, level_make,
                    scan_libs.
set_up:- ho_module(X, Y),
                    extract_name(X, NAME),
                    create_lib_list(NAME),
                    create_loaded, fail.
set_up:- write('Top level modules identified'),nl.
extract_name(X, N):- X=..Listname, prune_list(Listname, N).
prune_list([Head|Tail], Head).
create_lib_list(N):- not(member(N,[name_nil])),
retract(library_names(X)), app(X, [N], NN),
asserta(library_names(NN)).
```

```
create_loaded:- library_names(NN), asserta(loaded_lib_names(NN)),
asserta(dsl_level(NN,0)).
set_level(X):-retract(dsl_level(Y)), asserta(dsl_level(X)),
write(´DSL SIMULATOR: Observable level set to ´),write(X),nl.
load_dsl1:-
                    use_dsl_library(List),
                    level_pred,
                    process_lib(List).
load_dsl1:- write(´Finished loading libraries...´),nl.
level_pred:-!,
                    increment_level,
                    make_model_levels.
level_make:-
                    use_dsl_library(List),
                    update_count(List,-1), fail.
level_make:- write(´Built hierarchy tree...´),nl.
update_count(List, Newcount):-
                    dsl_level(X, Count),
                    not(dsl_level(List,XXX)),
                    Count > Newcount,
                    Count2 is Count + 1,
                    update_count(List, Count2).
update_count(List, N):-!,
                    not(dsl_level(List, XXX)),
                    N = \ = -1,
                    asserta(dsl_level(List, N)).
increment_level:-!,
                    dsl_level(X),
                    Y is X + 1,
                    retract(dsl_level(X)),
                    asserta(dsl_level(Y)).
make_model_levels:-
                    make_modules,
                    make_ho.
make_modules:-
                    module(X,_),
                    not(model_level(X, XX)),
                    dsl_level(L),
                    asserta(model_level(X, L)),fail.
make_modules:-
                    write(´updated module levels ´),nl.
make_ho:-
                    ho_module(X,_),
                    not(model_level(X, XX)),
                    dsl_level(L),
                    asserta(model_level(X, L)),fail.
make_ho:-
                    write(´updated ho levels...´),nl.
module(default, []),
model_level(default,-1).
scan_libs:- use_dsl_library(X), add_to_lib_names(X), fail.
scan_libs:- nl.
add_to_lib_names(New):- !, retract(library_names(List1)),
                            app(List1), New, Result),
                            asserta(library_names(Result)).
library_names([]).
/*
```

The *process_lib* predicate is quite complicated because of the level of backtracking involved. It should be noted that its first predicate will

eventually fail with an empty *list* = *List.* Hence backtracking is invoked to *load_dsl1* until all *use_dsl_library* predicates have been examined.
*/

```
process_lib(List):- get_first_element(List,First,Rest),
                    prune_lib_list(First),
                    process_lib(Rest).
```

/* Check to see if the *use_dsl* member is already loaded into the environment. If it is, do not load the library component in; however, *library_names* still maintains a record of reference to the duplicate call (to maintain the hierarchy).
*/

```
prune_lib_list(First):-
                    loaded_lib_names(LIST),
                    not(member(First, LIST)),
                    load_library(First),
                    write(' Loading library component '),
                    write(First),
                    write(' into current environment…'),nl,nl,
                    retract(loaded_lib_names(L)),
                    app(L, [First], NEW),
                    asserta(loaded_lib_names(NEW)).
```

/* If the duplicate is already in the environment, just ignore it.
*/

```
prune_lib_list(First):-
                    loaded_lib_names(LIST),
                    member(First, LIST),
                    remove_element(LIST, First, NEW),
                    app(NEW, [First], Newlist),
                    retract(loaded_lib_names(XXX)),
                    asserta(loaded_lib_names(Newlist)).
get_first_element([Head|Tail],Head,Tail).
load_library(DSL_NAME):- write('consulting library module '),
                    write(DSL_NAME),nl,
                    consult(DSL_NAME).
```

/***
get_index returns the index of a module name instance.
***/

```
get_index(Name, Value):- Name =.. Listname,
find_the_index(Listname, Value).
find_the_index([First, Value], Value).
```

/***/
The following predicate involves the build routines associated with libraries.
***/

```
lib_config:-
system('clear'),system('date'),nl,library_names(LIST),lib_config
(LIST).
lib_config([]):- write(' … All Library modules configured …
'),nl,nl.
lib_config([NAME|REST]):-
                        nl,write(' … Attempting to configure library
                        module '), write(NAME),nl,
                        configure_library(NAME),
                        dsl_level(LIST, LEVEL),
                        member(NAME, LIST),
                        Lev2 is LEVEL + 1,
                        retract(dsl_level(XXXX)),
                        asserta(dsl_level(Lev2)),
                        make_model_levels,
                        nl,
                        write('Configured '),
                        write(NAME),
                        write(' …\n'),
                        lib_config(REST).
```

/***

Simulated time : initial state

Essentially these are system variables placed into a default state. They are used by the simulator to schedule messages.

***/

```
sim_time(0.00).
current_module_processed(CURRENT).
max_sim_time(40000000.00).
use_dsl_library([]):-fail.
configure_library(nil).
isa(nil_default_setup_name, nil_parent_module_name).
constraint(nil_default, nil_message, nil_type, nil_value).
ho_module(name_nil,[]).
terminate(false).
```

/***

The module simulator routines:

A time of “-1” in the next time predicate indicates no messages scheduled. That is why we require a *check_res* on the sign of Temp below. The predicates in this block of code control the ordering of the information and message time delays. It is in fact a rudimentary simulator for DSL. A simulation time *sim_time(TIME)* emulates the virtual time.

***/

```
dsl:- write('DSL ENVIRONMENT Rev. 1.0 - 0295'),nl,
                        write('written by: Oryal Tanir, McGill
                        University MACS lab.'),nl,nl,
                        write('Extracting Library rules: '),nl,nl,
                        load_dsl,nl,nl,lib_config,nl,nl,
                        ho_module_lib_config,
                        write('DSL PROCESSOR: Setting up module
```

```
                         resources´),nl,initialize.
ho_module_lib_config:-
                         ho_module(X, []),
                         isa(X, OBJECT),
                         write(´Processing HO: ´), write(OBJECT),nl,
                         ho_module(OBJECT, SUBS),
                         process_ho_list(X, SUBS),
                         ho_path_config(X, OBJECT),
                         ho_module_lib_config.
ho_module_lib_config:-
                         write(´All HO- library modules instantiat-
                         ed.´),nl.
process_ho_list(X, PARAMETERS):-
                         retract(ho_module)X, [])),
                         asserta(ho_module(X, PARAMETERS)).
ho_path_config(X, OBJECT):-
                         path(OBJECT, D1, Z1),
                         asserta(path(X, D1, Z1)), fail.
ho_path_config(X, OBJECT):-
                         path(D1, OBJECT, Z1),
                         asserta(path(D1, X, Z1)), fail.
ho_path_config(X, OBJECT):-
                         write(´ Configuring ho_module paths…´),nl.
initialize:- module(Name,_), not(isa(_,Name)),
                         asserta(message_queue(Name, [])),
                         asserta(next_time(Name,-1)),
                         asserta(module_state(Name,ready)),
                         asserta(path(Name,Name,[in,out])),
                         fail.
initialize:- write(´DSL SIMULATOR: Module queues instantiated´),
nl,
                         write(´dsl simulator and processor
                         ready.´),nl,nl,nl,
                         write(´Type "start_cond_exec" to instantiate
                         start conditions and place system into a valid
                         starting state´),nl,
                         write(´Use step(X) for relative time-stepping
                         and ´),nl,
                         write(´start(X) for absolute time.´),nl,nl.
```

/***

The following is another simulation output format.

```
step:- sim_time(Main_time), system(´clear´),nl,
write(´############################################################´),
                         nl,
                         write(´DSL SIMULATOR: <<update>> CURRENT SIM-
                         ULATOR TIME is´),
                         write(Main_time),write(´ microseconds.´),nl,
write(´############################################################´),
                         nl,nl,
                         queue_handler,
```

***/

```
                         step:- sim_time(Main_time), system(´clear´),nl,
                         write(´—> CURRENT SIMULATOR TIME is ´),
                         write(Main_time),write(´ microseconds.´),nl,XT
                         is cputime,
                         write(´= = = = = = = = = = = = = = = = = = =
```

```
´),write(XT),
                          nl,nl,
                          queue_handler.
step:- end_sim.
step(0):- end_sim.
step(X):- step, XX is X-1, step(XX).
start:- sim_time(Main_time), system(´clear´),nl,
                          write(´—> CURRENT SIMULATOR TIME is ´),
                          write(Main_time),write(´ microseconds.´),nl, XT
                          is cputime.
                          write(´= = = = = = = = = = = = = = = = = = =
´),write(XT),
                          nl,nl,
                          repeat,
                          queue_handler.
                          not(manipulate_clock),
                          manipulate_clock2,test_sim_time.
start:- end_sim.
start(X):- retract(max_sim_time(_)), asserta(max_sim_time(X)),
start.
go:-XCPU is cputime, asserta(begin_time(XCPU)),repeat,start,termi-
nate(true), reset_streams.
end_sim:- nl,
                          write(´= = = = = = = Message from DSL
                          simulator = = = = = = = = = = = = ´),
                          nl,
                          write(´ No further messages to process: ´),nl,
                          write(´ Current simulated time is ´),
                          sim_time(T),
                          write(T),write(´ microsecs´),nl.
queue_handler:- sim_time(Main_time),
                          next_time(Modx, Main_time),
                          retract(current_module_processed(XMODULE)),
                          asserta(current_module_processed(Modx)),
                          message_queue(Modx,[(Message,Qtime)|Rest]),
                          isa_hierarchy(Modx, Module1),
                          isa_hierarchy(Message, Mess),
                          module(Module1, MethodList),
                          get_method(Mess, MethodList, Method),
                          next_time(Modx, Main_time),
                          test_and_exec(Method,Mess,Modx).
test_and_exec(Method,Mess,Modx):-
                          call(Method),!,
                          print_sim_time,
                          print(´DSL PROCESSOR: Message ´),print(Mess),
                          print(´ completed at module ´),printl(Modx),
                          check_constraint(Modx, Mess),
                          remove_first(Modx)
test_and_exec(Method,Mess,Modx):-
                          module_state(Modx, delayed).
debug_dsl:- list(n,s,10).
```

/***** queue_handler:- update_sim_clock. ****************************/

```
manipulate_clock:-
                          sim_time(Main),
                          next_time(Modx, Main).
manipulate_clock2:-
                          update_sim_clock.
update_sim_clock:- next_time(_, Time1),
                          max_sim_time(MaxTime),
```

```
                    sim_time(Main),
                    MaxTime> = Main,
                    Time1> = 0,
                    find_low_time(Time1, Next_time).
update_sim_clock:-
                    sim_time(Main_time), system('clear'),nl,
                    write('—> CURRENT SIMULATOR TIME is '),
                    write(Main_time),write(' microseconds.'),nl, XT
                    is cputime,
                    write('= = = = = = = = = = = = = = = = = =  '),
write(XT),
                    nl,nl.
test_sim_time:-
                    max_sim_time(MaxTime),
                    sim_time(Main),
                    (Main> = MaxTime,
                    retract(terminate(X)),
                    asserta(terminate(true)), end_sim; true).
find_low_time(Curr_time, Temp):- next_time(_, Temp),
                    Temp> = 0,
                    Curr_time>Temp,!,
                    find_low_time(Temp, Xt).
find_low_time(Curr_time, Curr_time):- retract(sim_time(Y)),
                    asserta(sim_time(Curr_time)).
remove_first(M):- module_state(M,ready),
                    retract(message_queue(M, [First|Rest])),
                    asserta(message_queue(M, Rest)),
                    null_check_res(M, Rest).
null_check_res(Module,[]):- retract(next_time(Module,_)),
                    asserta(next_time(Module,-1)).
null_check_res(Module,X).
```

/***

The following are some simulator predicates listing resources for the simulator state.

***/

```
list:- system('clear'),write('DSL SIMULATOR: Interactive List mode
\n'),
                    write('Do you wish to limit the scope of list
                    to only one level? (y/n) '),
                    get(S),error_check(S, SS),
                    char_code(MODE,SS),system('clear'),
                    write('\nPlease type an appropriate integer to
                    indicate level of observation: \n'),
                    get(L),char_code(LEV,
                    L),make_number(LEV,LEVEL),
                    write('Observation level for list is set to
                    '),write(LEVEL),nl,nl,
                    write('What do you wish to list?\n'),
                    write('r - resources\n'),
                    write('h - ho-modules\n'),
                    write('i - hierarchy\n'),
                    write('s - state of queues\n'),
                    write('p - path connections\n'),
                    write('q - quit'),
                    write('? '),
                    get(OP), char_code(OPTION,OP),system('clear'),
                    nl, list(MODE, OPTION, LEVEL).
```

```
error_check(121, 121).
error_check(89, 121).
error_check(78, 110).
error_check(110, 110).
error_check(X, 121).
list(X, q,L).
list(XX,i,L):-write('Hierarchy listing \n'),
                isa(X, Y),
                write(X), write(' \t\t -> '),
                write(Y),nl,fail.
list(y,r,L):-write('Listing Resources Created: \n'),
                write('LEVEL\tModule: \t\tResource:
                \t\t\tVariable(s): \t'),nl,nl,
                resource(NAME, RESNAME, PAR),
                otest_level(L_,NAME),
                write(NAME),write('\t\t'),write(RESNAME),
                write('\t\t\t'), write(PAR), nl,fail.
list(y,s, L):-write('Listing state of the module queues:\n'),
                write('LEVEL\tModule: \t\tMessages queued:
                \tDelayed until time:\n'),nl,
                message_queue(M, Q), otest_level(L,M),
                next_time(M, T),
                write(M),
                write(' \t'),
                write(Q),
                write('\t'),
                write(T),nl,
                fail.
list(y, h, L):- write('List of HO_Modules:'),nl,write('LEVEL\tHO-
MODULE NAME: \t\t [ Sub modules ]:\n'),nl,
                ho_module(X, Y), otest_level(L,X),
                write(X), write('\t'), write(Y), nl,fail.
list(y, p, L):-write('Listing Path interconnections:\n'),
                model_level(A,_),
                otest_level(L, A),
                write(A), write(' is connected to:\n'),
                list_the_paths(A).
list(n, p, L):-write('Listing Path interconnections:\n'),
                model_level(A,_),
                test_level(L, A),
                write(A), write(' is connected to:\n'),
                list_the_paths(A).
list_the_paths(A):-
                path(A,B,Y),
                write('\t -> '),write(B),write('\t Port pair:
                '),write(Y),nl,fail.
list(n, r,L):- write('Listing Resources Created:
\n'),write('LEVEL\tModule: \t\tResource: \t\t\tVariable(s):
\t'),nl,nl,
                resource(NAME, RESNAME, PAR),
                test_level(L,NAME),
                write(NAME),write('\t\t'),write(RESNAME),
                write('\t\t\t'), write(PAR), nl,fail.
list(n, s, L):- write('Listing state of the module queues:\n'),
                write('LEVEL\tModule: \t\tMessages queued:
                \tDelayed until time:\n'),nl,
                message_queue(M, Q), test_level(L,M),
                next_time(M, T),
                write(M),
                write(' \t'),
                write(Q),
                write('\t'),
```

```
                        write(T),nl,
                        fail.
list(n, h, L):- write('List of HO_Modules:'),nl,write('LEVEL\tHO-
MODULE NAME: \t\t [ Sub modules ]:\n'),nl,
                        ho_module(X, Y), test_level(L,X),
                        write(X), write('\t'), write(Y), nl,fail.
list(res):- list(r,10).
list(s):-list(s, 10).
list(h):-list(h, 10).
list:-nl,write('done.'),nl.
list(X, Y, Z):-nl,write('done.'),nl.
test_level(L,NAME):-
                        model_level(NAME, LEVEL),
                        L>=LEVEL,write(LEVEL), write('\t').
otest_level(L,NAME):-
                        model_level(NAME, LEVEL),
                        L=LEVEL,write(LEVEL), write('\t').
```

/***

Code for evaluating built-in constraints.

***/

```
check_constraint(Module, Mess):-
                                constraint(Module, Mess, Type,
                                Value),
                                print_sim_time,
                                test_constraint(Type, Module, Mess,
                                Value).
check_constraint(Module, Mess).
test_constraint(upper_limit, Module, Mess, Value):-
                        message_queue(Module,[(Mess,Qtime)|Rest]),
                        sim_time(Curr_time),
                        Queued_time is (Curr_time - Qtime),
                        Value<Queued_time,
                        write(' !!!!!!!!!!!!!!!!!!!!!!!!!!!!!!!!!!!!!!!!
                        '),nl,
                        print_sim_time,
                        write(' DSL SIMULATOR: WARNING: Upper_limit
                        Constraint Violation! '),
                        nl,
                        print_sim_time,
                        write(' Module: '),
                        write(Module),
                        write(' Message: '),
                        write(Mess),nl,
                        print_sim_time,
                        write(' Constrained Variable '),
                        write(Queued_time),
                        write(' exceeds upper limit of '),
                        write(Value),nl,
                        print_sim_time,
write(' !!!!!!!!!!!!!!!!!!!!!!!!!!!!!!!!!!!!!!!!'),nl.
test_constraint(upper_limit, Module, Mess, Value).
start_cond_exec:- start_cond, fail.
start_cond_exec:- nl,write('Initial simulator conditions set-up
complete…'),
                        nl.
start_cond:- nl, write('Initial simulator conditions being set-
up…').
                        :- system('clear'),system('date'),nl,nl,
```

```
                    write('DSL Interpreter is loaded into environ-
                    ment.'),
                    nl,
write('= = = = = = = = = = = = = = = = = = = = = = = = = = = ='),
                    nl,nl,
                    write(' If not done, consult your main DSL
                    model now and'),nl,
                    write(' type "run" to commence auto-configura-
                    tion, or for manual operation,'),
                    nl,
                    write(' use the following dsl interpreter com-
                    mands to configure environment:'),
                    nl,
                    write(' (Note: the order of commands should be
                    adhered to as below)'),
                    nl,
                    write('  "load_dsl, lib_config, initialize"
                    '),nl,
                    nl,
                    write(' load_dsl will access all necessary
                    library modules and resources,'),
                    nl,
                    write(' lib_config will configure all libraries
                    to your environment, and'),
                    nl,
                    write(' initialize will create necessary data
                    structures for the DSL simulator.'),
                    nl,nl.
write('= = = = = = = = = = = = = = = = = = = = = = = = = = = =),
                    nl,nl,
run:- dsl, system('rm outfile'), system('touch outfile'),
system('clear'), system('date'),nl,start_cond_exec.
synthesis:- synth, nl, write(' Synthesis initiated to VHDL:'),nl,
write('Pre-check performed on components'),nl,nl, scan_out,
write('Warning: not all DSL modules have defined types.'),nl,
write('Some modules may require extra attention in VHDL (marked
UNDEF)'),nl,nl,write('running parser'),nl,
system('sed -f sefile vhdl> v'), system('t<v>dslout'),
write('Synthesis complete').
synth:- module(Module, Meth), isa_hierarchy(Module, Module1),
not(module(Module1,[])), module(Module1, Method),
asserta(vhdl(Module, Method)), fail.
synth:- nl, write('DSL-VHDL SYNTHESIS————'),nl.
scan_out:- tell(vhdl), vhdl(M,Me), write('module('), write(M),
write(  ,  '), writeq(Me), write(').'), nl,fail.
scan_out:-told.
```

Appendix

B

DSL Specification of Telephone (DTSS) Example

```
ho_module(main(10, 0, 5, 2), [subscriber(X),
residential_switch(X)]).
/* Main parameters are: no of telephones, switch no, no of chan-
nels/card, no of int. cards */
use_dsl_library([
                    dts_switch
                    telephone
                    ]).
isa(subscriber(X), user(X)).
path(subscriber(T), residential_switch(SW), [tel_line(T),
tel_line(T)]).
configure_library(main):-
            ho_module(main(T, SW, CH, CARD), XXX),
            asserta(isa(residential_switch(SW), dts_switch(SW, T,
            CARD,
CH))),
            asserta(ho_module(residential_switch(SW), [])),
            configure_tel_con(T, SW).
configure_tel_con(1, SW):-
            asserta(module(subscriber(0), [])),
            write('Telephones connected to the switch...'),nl.
configure_tel_con(T, SW):-T>1,
            TT is T-1,
            asserta(module(subscriber(TT), [])),
            configure_tel_con(TT, SW).
start_cond:- retract(current_module_processed(XXX)),
            asserta(current_module_processed(subscriber(0))),
            send(start_a_call(9)).

= = = = = = = = = = = = = = = = = = = = = = = = = = = = = =

ho_module(dts_switch(NUMBER, INPUTS, CARDS, CHANNELS),
[switch_net(NUMBER),
interface_card(Y)]).
use_dsl_library([
                    generic_switch_element,
                    interface_comp
                    ]).
```

```
configure_library(dts_switch):-
                    isa(INSTANCE, dts_switch(N,I,C,CH)),
                    II is I//C,
                    asserta(isa(switch_net(N),
generic_switch_element(C,CH))),
                    asserta(isa(interface_card(Y),
interface_comp(II, CH))),
                    asserta(path(interface_card(Y), switch_net(N),
[int_port(Y), trunk(Y)])),
                    asserta(path(switch_net(N), interface_card(Y),
[trunk(Y), int_port(Y)])),
                    asserta(ho_module(switch_net(N), [])),
                    example_config(N, II, C).
example_config(N, IN, 1):-
                    Index is N*16,
                    asserta(ho_module(interface_card(Index),
                    [])),
                    tel_port_config(Index, IN, 0).
example_config(N, IN, CARD):-
                    CARD1 is CARD-1,
                    Index is N*16 + CARD1,
                    asserta(ho_module(interface_card(Index), [])),
                    tel_port_config(Index, IN, CARD1),
                    example_config(N, IN, CARD1).
tel_port_config(Index, 0, CARD1):-
                    nl.
tel_port_config(Index, IN, CARD1):-
                    isa(INSTANCE, dts_switch(N,I,C,CH)),
                    IPRIME is I//C,
                    INN is IN-1
                    Index2 is IPRIME*CARD1 + INN,
                    asserta(path(interface_card(Index), INSTANCE,
[twisted_pair(INN), tel_line(Index2)])),
                    tel_port_config(Index, INN, CARD1).

 = = = = = = = = = = = = = = = = = = = = = = = = = = = = = = =

/*------------- Time Switch Module ------------*/
module(generic_time_switch(Z),
            [
            (address_select(ADDRESS):- delay(0.1),
                    check_res(address_map(ADDRESS), CARD),
                    probe(switch_line(Z), DATA),
                    write('>>>>>>>>>>>>>>>>>>>>>>>>>> '),
                    write('Time_switch: latched onto data: '),
write(DATA),
                    send(_, line_port(CARD),
update_buffer(ADDRESS,DATA)) ),
            (address_select(ADD):-
                    write('>>>>>>>>>>>>>>>>>>>>>>>>>>
Time_switch: Address '),
                    write(ADD),write(' undefined.'),nl)
            ]).
/* Library resources */
lib_resource( generic_time_switch,[address_map(X)]).
lib_spec_resource(generic_time_switch,input_ports, [proc_port]).
lib_spec_resource(generic_time_switch,output_ports, [switch_line,
line_port(N)]).
/* Library path declarations. - none for generic libraries*/
/* build rules:
The library module will automatically build the necessary
resources.
In this case these are the channel out buffer allocations.*/
```

```
configure_library(generic_time_switch):-
            isa(X,generic_time_switch(XX)),
            isa(NAME, generic_lineout(_)),
            module(X,_),
            get_index(X, I),
            Low_bound is (I*16-1),
            Up_bound is (I*16 + 16),
            module(NAME,_),
            get_index(NAME,Index),
            Low_bound<Index,
            Up_bound>Index,
            resource(NAME, channel_out_buffer(Value), LL),
            asserta(resource(X, address_map(Value), Index)),
            fail.
configure_library(generic_time_switch):-
            write(´ generic_time_switch: All resources successfully
configured...´),nl.
constraint(time_switch(Z), address_select(A), upper_limit, 0.11).

= = = = = = = = = = = = = = = = = = = = = = = = = = = = = =

/*------------- Frame Control Module ------------*/
module(generic_frame_control(Z, LIMIT),
            [
            (frame_start:- check_res(switch_frame_counter, COUNT),
                           check_res(number_of_in_cards,
INPUT),
                           COUNT<LIMIT,
                           TEMP is LIMIT*INPUT,
                           FDELAY is 125/TEMP,
                           delay(FDELAY),
                           COUNT2 is COUNT + 1,
write(´>>>>>>>>>>>>>>>>>>>>>>>>   ´),
                           write(´FRAME CONTROL: new frame
count =  ´),
                           write(COUNT2),nl,
                           set_res(switch_frame_counter,
COUNT2),
                           send(channel_detect(COUNT2,
INPUT))),
            (channel_detect(COUNT, INPUT):-
                           COUNT<INPUT,
                           send(sp,(Z), sp_sync_port(Z),
command_frame_select(COUNT)),
                           send(control_processor(Z),
proc_in_port(Z), command_frame_select(COUNT)),
                           send(frame_start) ),
            (channel_detect(COUNT, INPUT):-
                           send(sp(Z), sp_sync_port(Z),
frame_select(COUNT)),
                           send(control_processor(Z),
proc_in_port(Z), frame_select(COUNT)),
                           send(frame_start) ),
            (frame_start:-check_res(switch_frame_counter,COUNT),
                           LIMIT2 is LIMIT-1,
                           COUNT>LIMIT2,
                           set_res(switch_frame_counter, 0),
                           send(frame_start) )
            ]).
lib_resource(
generic_generic_frame_control,[frame_counter,frame_count_limit,num-
ber_of_in
_cards]).
```

```
lib_spec_resource(generic_generic_frame_control,input_ports, []).
lib_spec_resource(generic_generic_frame_control,output_ports,
[proc_in_port,sp_sync_port]).
/* build rules:
The library module will automatically build the necessary
resources.
In this case, these are the channel out buffer allocations.*/
configure_library(generic_frame_control):-
            isa(X,generic_frame_control(ZZ,_)),
            module(X, List),
            get_index(X, Z),
            asserta(resource(frame_control(Z), switch_frame_
            counter, 0)),
            generic_frame_control1(Z),
            fail.
configure_library(generic_frame_control):- write(´
generic_frame_control:
All resources successfully configured…´),nl.
generic_frame_control1(Z):-
            asserta(resource(frame_control(Z), number_of_in_cards,
            0)),
            isa(X, generic_linein(XX)),
            module(X,_),
            resource(frame_control(Z), number_of_in_cards, COUNT),
            COUNT1 is (COUNT + 1),
            retract(resource(frame_control(Z), number_of_in_cards,
            XXX)),
            asserta(resource(frame_control(Z), number_of_in_cards,
COUNT1)),
            write(´ generic_frame_control: Number of line cards
            detected
is ´),
            write(COUNT1),nl,
            fail.
```

= =

```
/*————————- Serial/Parallel module —————*/
module(generic_sp(X),
    [
    (buffer_update(ADDRESS,DATA):- delay(1.5),
                set_res(channel_in_buffer(ADDRESS), DATA),
                write(´>>>>>>>>>>>>>>>>>>>>>>>>>´),
                write(´SP MODULE: In_buffer updated for channel
´),
                write(ADDRESS),nl ),
    (frame_select(COUNT):- check_res(channel_in_buffer(COUNT),
    VOICE),
                delay(0.8),
                check_res(channel_in_buffer(COUNT), VOICE),
send(switch_bus(X),sp_port(X),switch_data(VOICE))),
    (frame_select(COUNT):- nl,write(´>>>>>>>>>>>>>>>>>>´),
                write(´SP module: switch bus unused at frame
                ´),
                write(COUNT),nl),
    (command_frame_select(Channel):-
                check_res(channel_in_buffer(Channel), COMMAND),
                send(_, proc_in_port(X),
channel_command(Channel,COMMAND)))
    ]).
/* Library resources */
lib_resource( generic_sp,[channel_in_buffer(X)]).
```

```
lib_spec_resource(generic_sp,input_ports, [fame_sync,ds_port]).
lib_spec_resource(generic_sp,output_ports, [sp_port]).
/* Library path declarations—none for generic libraries*/
/* build rules:
The library module will automatically build the necessary
resources. In this case these are the channel out buffer alloca-
tions.*/
configure_library(generic_sp):-
            isa(NAME, generic_sp(I)),
            isa(DSNAME, generic_linein(XX)),
            module(NAME,_),
            get_index(NAME, Index),
            module(DSNAME,_),
            get_index(DSNAME, Value),
            isa(NAME1, generic_switch_element(NNLINES,MMCHAN)),
            MM is MMCHAN-1,
            Low_bound is (Index*16-1),
            Up_bound is (Index*16 + 16),
            Low_bound<Value,
            Up_bound>Value,
            generic_sp1(Index,Value,MM),
            fail.
configure_library(generic_sp):-
            write(´ generic sp: All resources successfully
configured...´),nl.
generic_sp1(NO, Y, -1):- write(´ generic_sp1: Resource
channel_in_buffer for
SP has been configured...´),nl.
generic_sp1(NO, Index,COUNT):- COUNT>-1, TEMP is (COUNT*20 +
Index),
            asserta(resource(sp(NO), channel_in_buffer(TEMP),
lineindata)),
            Next is (COUNT-1),
            generic_sp1(NO,Index, Next).

= = = = = = = = = = = = = = = = = = = = = = = = = = = = = = =

/* ————————- Generic - clock unit ————————
    This module provides a modulo-NO_OF_CHANNEL counter for the
    line
    output modules. We assume that every channel within the
    line protocol is 10 bits wide, with a "sync" bit sent
    at the beginning of each transmission. A clock(COUNT) is sent
    to the destination at a constant "BITS." A frame_cycle
    message completes the clock circuitry to indicate a completion
    of the module count and commence a sync transmit */
module(generic_clock(INDEX, NO_OF_CHANNEL),
    [
    (clock_count(NO_OF_CHANNEL):- send(clock_count(0)),
                    send(_,clock_port(INDEX),frame_cycle)),
    (clock_count(NEW):- NEW<NO_OF_CHANNEL,
                    Clock_rate is (125/
NO_OF_CHANNEL),
                    delay(Clock_rate),
                    COUNT is NEW + 1,
                    send(_,clock_port(NO),clock(NEW)),
                    send(clock_count(COUNT)))
    ]).
start_cond:- write(´Setting up initial messages ´),nl,
            isa(X, generic_clock(_, NO)),
            retract(current_module_processed(XXX)),
```

```
            asserta(current_module_processed(X)),
            send(clock_count(NO)).
configure_library(generic_clock).

= = = = = = = = = = = = = = = = = = = = = = = = = = = = = =

/*-------------- line line card module (in) ----------*/
module(generic_linein(Z),
    [
    (channel_sync:- set_res(channel_count,0),
                    write('>>>>>>>>>>>>>>>>>>>>>>>>>'),
                    write('line line in card:channel sync
received'),nl,
    (channel_message(DATA):- check_res(channel_count,CHANNEL),
    check_res(card_number, CARD),
    CHAN is (CHANNEL*20 + CARD),
    write('>>>>>>>>>>>>>>>>>>>>>>>>>'),
    write('line line in card: updating channel '),
    write(CHANNEL),write(' with data:
')write(DATA),nl,
                    CHAN2 is CHANNEL + 1,
                    set_res(channel_count,CHAN2),
                    send(_, sp_port(Z), buffer_update(CHAN,DATA)) )
    ]).
/* Library resources */
lib_resource( generic_linein, [card_number,channel_count]).
lib_spec_resource(generic_linein, input_ports, [in_port]).
lib_spec_resource(generic_linein, output_ports, [sp_port]).
/* Library path declarations-none for generic libraries*/
/* build rules:
The library module will automatically build the necessary
resources.
In this case, these are the channel out buffer allocations.*/
configure_library(generic_linein):-
            isa(X,generic_linein(_)),
            module(X, List),
            get_index(X, Value),
            asserta(resource(X, card_number, Value)),
            asserta(resource(X, channel_count,0)),
            fail.
configure_library(generic_linein):-
            write(' generic_linein: All resources successfully
configured...'),nl.

= = = = = = = = = = = = = = = = = = = = = = = = = = = = = =

/*
LIBRARY MODULE:     line line card module (out).
This is a generic line output line card description. This library
module automatically configures instances invoked in the main DSL
program. Its configuration affects the resource variables
card_number
channel_out_buffer
*/
module(generic_lineout(Z),
    [
    (frame_cycle:-
                    send(_,out_port(Z),channel_sync),
                    write('>>>>>>>>>>>>>>>>>>>>>>>>
                    '),
                    write('line output card: Frame sync message
                    sent
```

```
´),nl),
     (clock(0):-
                    check_res(card_number, CARD),
                    check_res(channel_out_buffer(CARD), MESSAGE),
                    send(_,out_port(Z),channel_message(MESSAGE)),
                    write(´>>>>>>>>>>>>>>>>>>>>>>>>>
                    ´),
                    write(´Line out card number: ´), write(CARD),
                    write(´ on channel 0 OUTPUTTED command ´),
                    write(MESSAGE),
                    nl),
                    set_res(channel_out_buffer(CARD), nop),
     (clock(COUNT):-
                    check_res(card_number, CARD),
                    COUNT2 is (COUNT*20 + CARD),
                    check_res(channel_out_buffer(COUNT2), VOICE),
                    send(_,out_port(Z),channel_message(VOICE)),
                    write(´>>>>>>>>>>>>>>>>>>>>>>>>   ´),
                    write(´line out card number: ´), write(CARD),
                    write(´ : OUTPUTTED message ´),
                    write (VOICE),
                    write(´ in channel ´),
                    write(COUNT),nl),
     (update_buffer(ADD,DATA):- ADD>15,
          set_res(channel_out_buffer(ADD), DATA),
          write(´>>>>>>>>>>>>>>>>>>>>>>>>>   ´),
          write(´line out card: new channel data updated´), nl,
          write(´ address is ´), write(ADD),nl,nl),
     (update_buffer(ADD,DATA):-
          check_res(channel_out_buffer(ADD),nop),
          set_res(channel_out_buffer(ADD), DATA),
          write(´>>>>>>>>>>>>>>>>>>>>>>>>>   ´),
          write(´line out card: new channel data updated´), nl,
          write(´ address is ´), write(ADD),nl,nl),
     (update_buffer(ADD,DATA):-
          delay(0.5),
          end(update_buffer(ADD,DATA)))
     ]).
/* Library resources */
lib_resource( generic_lineout,[card_number,channel_out_buffer(X)]).
lib_spec_resource(generic_lineout,input_ports, [dscp(Number),
ts_line_port(Number)]).
lib_spec_resource(generic_lineout,output_ports,
[out_port(Number)]).
/* Library path declarations—none for generic libraries*/
/* build rules:
The library module will automatically build the necessary
resources.
In this case, these are the channel out buffer allocations.*/
configure_library(generic_lineout):-
          isa(X,generic_lineout(Z)),
          module(X, List),
          get_index(X, Value),
          asserta(resource(X, card_number, Value)),
          generic_lineout1(X,Value,0),
          fail.
configure_library(generic_lineout):-
          write(´ generic_lineout: All resources successfully
configured…´),nl.
generic_lineout1(X,Y, 32):- nl,
          write(´ generic_lineout: Resource channel_out_buffer
          for ´),
```

```
            write(X),write(' has been configured…'),nl.
generic_lineout1(Name,Index,COUNT):- COUNT<32, TEMP is (COUNT*20 +
Index),
            asserta(resource(Name, channel_out_buffer(TEMP),
lineoutdata)),
            Increment is (COUNT + 1),
            generic_lineout1(Name, Index, Increment).
/* Automatic startup of configuration enabled below*/

= = = = = = = = = = = = = = = = = = = = = = = = = = = = = = = = = =

ho_module(interface_comp(INPUT, CH),
                    [
                    line_card(CARD), controller(CARD), clock(CARD)
                    ]).
use_dsl_library([
                    sub_server,
                    line_card,
                    generic_clock
                    ]).
configure_library(interface_comp):-
                    isa(X, interface_comp(INPUT,CH)),
                    ho_module(X, YYY),
                    get_index(X,CARD),
                    asserta(module(controller(CARD), [])),
                    asserta(module(line_interface(CARD), [])),
                    asserta(module(timer(CARD), [])),
                    asserta(path(X, controller(CARD),
[twisted_pair(Y), twisted_pair(Y)])),
                    asserta(path(controller(CARD), X,
[twisted_pair(Y), twisted_pair(Y)])),
                    asserta(path(X, line_interface(CARD),
[int_port(CARD), out_port(CARD)])),
                    asserta(path(line_interface(CARD), X,
[out_port(CARD), int_port(CARD)])),
                    asserta(path(line_interface(CARD),
controller(CARD), [proc_port(CARD), proc_port(CARD)])),
                    asserta(path(controller(CARD),
line_interface(CARD), [proc_port(CARD), proc_port(CARD)])),
                    asserta(path(timer(CARD), line_interface(CARD),
[clock_port(CARD), clock_port(CARD)])),
                    asserta(isa(line_interface(CARD),
line_card(CARD, CH))),
                    asserta(isa(timer(CARD), generic_clock(CARD,
CH))),
                    asserta(isa(controller(CARD), sub_server(CARD,
INPUT))),
                    fail.
configure_library(interface_comp):- nl,nl, write(' Completed con-
figuring
interface components.'),nl.

= = = = = = = = = = = = = = = = = = = = = = = = = = = = = = = = = =

/*
LIBRARY MODULE:     line card module (out).
This is a generic line output line card description. This library
module automatically configures instances invoked in the main DSL
program. The configuration affects the resource variables
card_number
channel_out_buffer
*/
```

```
module(line_card(Z, INPUT),
     [
     (frame_cycle:-
                    send(_,out_port(Z),channel_sync),
                    write('>>>>>>>>>>>>>>>>>>>>>>>>>
                    '),
                    write(' line output card: Frame sync message
                    sent
'),nl),
     (clock(COUNT):-
                    check_res(channel_buffer(COUNT), VOICE),
                    send(_,out_port(Z),channel_message(VOICE)),
                    write('>>>>>>>>>>>>>>>>>>>>>>>>>
                    '),
                    write(' line out: OUTPUTTED message '),
                    write(VOICE),
                    write(' in channel '),
                    write(COUNT),nl),
     (update_buffer(ADD,DATA):-
     check_res(channel_out_buffer(ADD),XX),
                    set_res(channel_buffer(ADD), DATA)),
     (channel_sync:- set_res(channel_count,0),
                    write('>>>>>>>>>>>>>>>>>>>>>>>>>
                    '),
                    write('line card:channel sync received'),nl),
     (channel_message(DATA):- check_res(channel_count,CHANNEL),
                    CHAN2 is CHANNEL + 1,
                    set_res(channel_count,CHAN2),
                    send(_, proc_port(Z),
channel_update(CHANNEL,DATA)) )
     ]).
/* build rules:
The library module will automatically build the necessary
resources.
In this case, these are the channel out buffer allocations.*/
configure_library(line_card):-
            isa(X,line_card(Z, CH)),
            module(X, List),
            get_index(X, Value),
            configure_buffers(X,Value,CH),
            fail.
configure_library(line_card):-
            write(' generic_out: All resources successfully
configured…'),nl.
configure_buffers(X,Y, -1):- nl,
            write(' line_card: Resource channel out buffer for '),
            write(X),write(' has been configured…'),nl.
configure_buffers(Name,Index,COUNT):- COUNT>-1,
            asserta(resource(Name, channel_out_buffer(COUNT),
            unused)),
            Decrement is (COUNT-1),
            configure_buffers(Name, Index, Decrement).

= = = = = = = = = = = = = = = = = = = = = = = = = = = = =

/* POTS SERVER (Plain Old Telephone Service) behavior is given
below.
This state descriptor encapsulates most features of call-processing
software on a switch.
TEL, REC = switch no, ds no, local no.
    i.e., 1, 4, 4567 makes up a phone number.
*/
```

```
module(pots_server(ID),
[
(responded_call(telno(X,DSB,B)):-
                        check_res(connect(A, telno(X,DSB,B)), ringing),
                        check_res(terminator(B), Bport),
                        set_res(connect(A,telno(X,DSB,B)), connected),
                        send(_,line_port(Bport),
update_buffer(Bport,change_state(B, busy)))
),
(call_dialed (telno(ID, SDNO, STEL), telno(RSNO, RDNO, RTEL)):-
                        create(connect(telno(ID, SDNO, STEL),
telno(RSNO, RDNO, RTEL)), connect_request),
                        create(agent(STEL), SDNO),
                        create(terminator(RTEL), RDNO),
                        send(_,line_port(RDNO), update_buffer(RDNO,
line_status(RTEL)))
),
(status_response(B, STATE):-
                        check_res(connect(A, B), connect_request),
                        send(called_state(A, B, STATE).
),
(phone_on_hook(telno(X, DS,A)):-
                        check_res(connect(telno(X, DS,A),telno(Y,
DSB,B)), connected),
                        remove(connect(telno(X, DS,A), telno(Y,
DSB,B))),
                        remove(agent(A), Aport),
                        remove(terminator(B), Bport),
                        set_res(channel_table(Aport, ChA, unused),
                        set_res(channel_table(Bport, ChB, unused),
                        send(_,line_port(Bport),
update_buffer(Bport,phone_off_hook(B)))
),
(phone_on_hook(B):-
                        check_res(connect(A,B), connected)
),
(phone_on_hook(telno(SW,DS,A)):-
                        check_res(agent(A), Aport),
                        remove(agent(A), Aport),
                        send(_,line_port(Aport), update_buffer(Aport,
change_state(A, onhook)))
),
(phone_on_hook(telno(SW,Aport,A)):-
                        send(_,line_port(Aport), update_buffer(Aport,
change_state(A, onhook)))
),
(called_state(telno(SW,DS,A), telno(BSW,BDS,B), busy):-
                        check_res(agent(A), Aport),
                        remove(connect(SW, telno(DS,A), telno(BSW,
BDS,B)), SS),
                        remove(terminator(B), Bport),
                        send(_,line_port(Aport), update_buffer(Aport,
busy_tone(A)))
),
(called_state(telno(XX,DS,A), telno(YY, BDS,B), onhook):-
                        check_res(agent(A), Aport),
                        check_res(terminator(B), Bport),
                        check_res(channel_table(Aport, ChA), unused),
                        set_res(channel_table(Aport, ChA), A),
                        check_res(channel_table(Bport, ChB), unused),
                        set_res(channel_table(Bport, ChB), B),
                        Index1 is ChA*20 + Aport,
                        Index2 is ChB*20 + Bport,
```

```
                         set_res(connect(telno(XX,DS,A), telno(YY,
BDS,B)), ringing),
                         send(_,mem_port(ID), mem_write(Index1,
Index2)),
                         send(_,mem_port(ID), mem_write(Index2,
Index1)),
                         send(_,line_port(Aport), update_buffer(Aport,
channel_allocate(A, ChA))),
                         send(_,line_port(Bport),
update_buffer(Bport,channel_allocate(B, ChB))),
                         send(_,line_port(Bport),
update_buffer(Bport,change_state(B, ringing))),
                         send(_,line_port(Aport), update_buffer(Index1,
ring_tone)),
                         send(_,line_port(Bport), update_buffer(Index2,
ring_tone))
),
(called_state(telno(XX,DS,A), telno(YY, BDS,B), onhook):-
                         write('No free lines available'),
                         remove(connect(telno(XX,DS,A), BB), SS),
                         remove(terminator(B), Bport),
                         send(_,line_port(DS), update_buffer(DS,
fast_busy(A)))
)
]).
configure_library(pots_server).

= = = = = = = = = = = = = = = = = = = = = = = = = = = = = =

module(sub_server(ID, INPUT),
[
(channel_update(0, COMMAND):-send (COMMAND)
),
(channel_update(CH, unused):- write('ignored data'),nl
),
(channel_update(CH, VOICE):- check_res(connect_table(TEL), CH),
                         send(_, twisted_pair(TEL), VOICE)
),
(phone_off_hook(A):-
                         check_res(line_state(A), CALLSTATE),
                         send(resolve_call_state(A, CALLSTATE))
),
(resolve_call_state(A, onhook):-
                         set_res(line_state(A), offhook),
                         send(_, twisted_pair(A), dialtone)
),
(resolve_call_state(A, busy):-
                         set_res(line_state(A), offhook),
                         send(_, twisted_pair(A), dialtone)),
(resolve_call_state(B, ringing):-
                         send(_, proc_port(ID),
update_buffer(0,responded_call(telno(X,ID,B)) ))
),
(change_state(TEL, STATE):-
                         set_res(line_state(TEL), STATE)
),
(dial_dest(A, B):-
                         send(_, proc_port(ID),
update_buffer(0, call_dialed(telno(X,ID,A), B)))
),
(phone_inactive(A):-
                         set_res(line_state(A), onhook),
                         send(_, proc_port(ID),
```

```
update_buffer(0,phone_on_hook(telno(X,ID,A) )))
),
(line_status(A):-
                    check_res(line_state(A), STATE),
                    send(_,proc_port(ID),
update_buffer(0,status_response(telno(X,ID,A),STATE)))
),
(channel_allocate(TEL, CH):-
                    set_res(connect_table(TEL), CH)
),
(fast_busy(A):- send (_, twisted_pair(A), fastbusy)
),
(busy_tone(A):- send (_, twisted_pair(A), busytone)
),
(voice_message(A, VOICE):-
                    check_res(connect_table(A), CH),
                    send(_, proc_port(ID), update_buffer(CH,
voice_message(VOICE)) )
)
]).
configure_library(sub_server):-
                    isa(X, sub_server(ID, IN)),
                    module(X, Y),
                    get_index(X, Index),
                    configure_subscribers(X,IN).
configure_subscribers(X, 0):-
                    write(´configured subscribers for interface
                    card
´),
                    write(X),nl.
configure_subscribers(X,IN):-IN>0,
                    asserta(resource(X, line_state(IN), on_hook)),
                    asserta(resource(X, connect_table(IN),
unused)),
                    IN_NEXT is IN -1,
                    configure_subscribers(X,IN_NEXT).

 = = = = = = = = = = = = = = = = = = = = = = = = = = = = = =

/*outputs
voice_message(Tel, VOICE)
dial_dest(TEL, DEST)
phone_off_hook(TEL)
phone_inactive(TEL)
inputs
voice_message(VOICE)
dialtone
fastbusy
busytone
ring_tone
*/
module(telephone(TEL),
[
(lift_handset:- send(_, tel_line(TEL), phone_off_hook(TEL))
),
(close:-send(_, tel_line(TEL), phone_inactive(TEL))
),
(digits(DEST) :- send(_, tel_line(TEL), dial_dest(TEL, DEST))
),
(dialtone:- send(_, headset(TEL), dialtone)
),
(fastbusy:-send(_, headset(TEL), fasttone)
```

```
),
(ring_tone:- send(_, headset(TEL), ringtone)
),
(busytone:- send(_, headset(TEL), busytone)
),
(voice_message(M):-send(_, headset(TEL), voice(M))
),
(send_voice(M):- send(_, tel_line(TEL), voice_message(M))
)
]).
module(user(TEL),
[
(start_a_call(DEST):- create(dest, DEST),
                      create(patience, 10),
                      send(_, hand(TEL), lift_handset)
),
(dialtone:- check_res(dest, DEST),
                      send(_, hand(TEL), digits(DEST))
),
(fastbusy:- send(test_patience(0))
),
(busytone:- send(test_patience(0))
),
(ringtone:- check_res(patience, V), VV is V-1,
                      send(test_patience(VV))
),
(voice(M):- print('voice call on headset'),nl
),
(talk(M):- send(_, hand(TEL), send_voice(M))
),
(test_patience(0):- remove(dest, DEST),
                      remove(patience, V),
                      send(_, hand(TEL), close)
)
]).
path(user(X), telephone(X), [hand(X), headset(X)]).
path(telephone(X), user(X), [headset(X), hand(X)]).
configure_library(telephone).
```

Appendix

Yacc Listing for Parser

The following is the Yacc listing for the DSL-VHDL parser.

```
%{
#include<string.h>
#include<stdio.h>
/* Let us first define our structure fields:
if name = entity
     We interpret this node as containing the entity name; the data
     field would contain the parameters.
     ptr1 would point to the next entity type node.
     ptr2 would point to the process type nodes associated with this
     entity.
     ptr3, ptr4 are NULL.

if name = process
     The data field would track the parameters.
     ptr1 would point to the next process.
     ptr2 would point to statements within this process.
     ptr3 would point to signals associated with the process.
     ptr4 would point to type determined for local variables.

if name = statement
     The data field would indicate a condition for the "command"
     fields to be executed.
     ptr1 would point to the next statement.
     ptr2 would point to relevant command nodes.
     ptr3 would point to condition nodes.
     ptr4 is NULL.

if name = command type
     the data1 field would indicate a VHDL command line form.
     data is the resource applied to.
     ptr1 would point to next command node.
     ptr2 would point to a destination field node.
     ptr3, ptr4 are NULL.

if name = signal
     The data field would possess the signal name.
     ptr1 would point to next signal node.
     ptr2, ptr3, ptr4 are NULL.
```

```
if name = destination
    The data field indicates the destination to be sent to.
    All other ptrs are NULL.

finally, the conditions node:
    name = condition
       data = value
       data1 = type

*/
static struct node {
    char name[400];/* type of node, i.e., entity name, process,
    statement, signal, or command */
    char data[500];/* relevant data are context dependent*/
    char data1[500];
    struct node *ptr1;/* Pointer declarations */
    struct node *ptr2;
    struct node *ptr3;
    struct node *ptr4;
};

static struct node *entity_ptr = NULL;/* pointer to current entity
processed */
static struct node *head_ptr = NULL;/* pointer to topmost entity */
static struct node *new_ptr; /* pointer for new node structs */
static struct node *proc_ptr, *stat_ptr;
static struct node *current_ptr;
static struct node *temp_ptr, *another_ptr;

static char scratch[400]; /* scratch buffer */
static char resource_name[400];/* temporary holder of resource
names */
static char delay_buffer[50];
static char current_command;
static int count, icount, dsl_state, send_state = 0;
static int measure;

/*static enum command_type
{UNDEF, SET, CREATE, REMOVE, DELAY, PROBE, WRITE, SEND, CHECK}
current_command; temporary holder of dsl command
*/
%}
%union{
     int i;
     char *text;
}
%token DSL_NAME
%token PORT_NAME
%token CONDITIONAL_OPERATOR
%token ENDER
%token GOAL
%token SETRES
%token CHECKRES
%token DELAY
%token PROBE
%token SEND
%token WRITE
%token DSL_VAR
%token MODULE
%token NAME
%token OPERATOR
%token INTEGER
```

```
%token PARAM
%token HOMODULE
%token CREATE
%token REMOVE
%token COMMENTS

%%
dsl_code: module_definition
     | ho_definition
     | dsl_code module_definition
     | dsl_code ho_definition
     ;
module_definition:
     MODULE ´(´ module_name ´,´ module_body
     {
     printf(¨ \n\n\nCOMPLETED A FULL MODULE PARSE\n¨);
     temp_ptr = entity_ptr;
     printf(¨ENTITY [data:] %s \n¨, temp_ptr->data);
     proc_ptr = temp_ptr->ptr2;
        while (temp_ptr ! = NULL)
         {
         printf(¨ = = [name:] %s [data:] %s \n¨, proc_ptr->name,
         proc_ptr->data);
         temp_ptr = proc_ptr->ptr4;
         while (temp_ptr ! = NULL)
          {
          printf(¨ VARIABLES [name:] %s [data:] %s \n¨, temp_ptr-
          >name, temp_ptr->data);
          temp_ptr = temp_ptr->ptr1;
          }

     temp_ptr = proc_ptr->ptr3;
     while (temp_ptr ! = NULL)
        {
        printf(¨ [name:] %s [data:] %s\n¨, temp_ptr-name, temp_ptr-
        >data, temp_ptr->data1);
     temp_ptr = temp_ptr->ptr1;
     }

     temp_ptr = proc_ptr->ptr2;
     while (temp_ptr ! = NULL)
        {/*Printing statements*/
     printf(¨  [name:] %s [data:] %s\n¨, temp_ptr->name,temp_ptr-
     >data);
        current_ptr = temp_ptr->ptr3;
        while(current_ptr ! = NULL)
        {/*Printing conditions*/
        printf(¨  [data1:] %s [data:] %s\n¨, current_ptr->data1,
        current_ptr->data);
        current_ptr = current_ptr->ptr1;
        }

        current_ptr = temp_ptr->ptr2;
        while(current_ptr ! = NULL)
        {/*Printing commands*/
        printf(¨* [name:] %s [data:] %s [data1:] %s\n¨, current_ptr-
        >name, current_ptr->data, current_ptr->data1);

         another_ptr = current_ptr->ptr2;
         while(another_ptr ! = NULL)
         { printf(¨PTR2_PAR: [name:] %s [data:] %s [data1:] %s\n¨,
         another_ptr->name, another_ptr->data,
```

```
another_ptr->data1);
         another_ptr = another_ptr->ptr1;
         }
         another_ptr = current_ptr->ptr3;
         while(another_ptr ! = NULL)
         { printf("PTR3_PAR: [name:] %s [data:] %s [data1:] %s\n",
another_ptr->name, another_ptr->data,
another_ptr->data1);
         another_ptr = another_ptr->ptr1;
         }

     current_ptr = current_ptr->ptr1;
     }
        temp_ptr = temp_ptr->ptr1;
        }
        printf("\n\n");
        proc_ptr = proc_ptr->ptr1;
        }

     }
     ;

module_body:'[' behavior_block ENDER
     | '[' ENDER
     {printf("YACC: DONE SCANNING \n");}
     ;

module_name:DSL_NAME
     {new_ptr = (struct node *)malloc(sizeof(struct node));

     (void) strcpy((*new_ptr).name, "ENTITY");
     (void) strcpy((*new_ptr).data, yytext);
     new_ptr->ptr1 = NULL;
     new_ptr->ptr2 = NULL;
     new_ptr->ptr3 = NULL;
     new_ptr->ptr4 = NULL;

     if (head_ptr = = NULL)
        {head_ptr = new_ptr;
        entity_ptr = new_ptr;
        }
     else
        {entity_ptr->ptr1 = new_ptr;
        entity_ptr = new_ptr;
        }
        proc_ptr = entity_ptr;
        }
        | module_name '(' module_param ')'
        ;

module_param:
     DSL_VAR
     | INTEGER
     |DSL_VAR ',' module_param
     | INTEGER ',' module_param
     ;

behaviour_block:behaviour_desc
     | behaviour_block ',' behaviour_desc
     ;

behaviour_desc:'(' behaviour_header GOAL behaviour_statement ')'
```

```
     ;

behaviour_header:DSL_NAME
     {

     /* Below checks for the occurrence of DSLNAME in process_list.
     The occurrence of DSLNAME indicates an option in VHDL (if-then-
     else-…); hence the data structure will generate a new state-
     ments list to hold the conditions, as well as the possible com-
     mand lists that will follow. In either case, the command
     pointer is set to null, indicating that it is expecting the
     first command addressed through stat_ptr.

     If no DSLNAME exists, then a new process list is malloced and
     inserted, as well as all pointers being updated.

     */

        current_ptr = proc_ptr;
        if ( strcmp(proc_ptr->name,"ENTITY") = = 0)
         {
         /* This indicates that there are no other processes. Hence
         we first set up the process (filled) and a statement
         (unfilled) node, which will get filled by the next parame-
         ters passed from lex. */

         new_ptr = (struct node *)malloc(sizeof(struct node));
         (void)strcpy(new_ptr->name, "PROCESS");
         (void)strcpy(new_ptr->data, yytext);
         new_ptr->ptr1 = NULL;
         new_ptr->ptr2 = NULL;
         new_ptr->ptr3 = NULL;
         new_ptr->ptr4 = NULL;

         proc_ptr->ptr2 = new_ptr;
         proc_ptr = new_ptr;
         current_ptr = proc_ptr;

         /*printf("REGISTERED A PROCESS NAME %s\n", proc_ptr-
         >data);*/

         new_ptr = (struct node *)malloc(sizeof(struct node));
         (void)strcpy(new_ptr->name, "STATEMENT");
         (void)strcpy(new_ptr->data, "UNDEF");
         new_ptr->ptr1 = NULL;
         new_ptr->ptr2 = NULL;
         new_ptr->ptr3 = NULL;
         new_ptr->ptr4 = NULL;
         current_ptr->ptr2 = new_ptr;
         current_ptr = new_ptr;

         printf("YACC:REGISTERED A STATEMENT FOR PROC %s\n",
         proc_ptr->data);
         }
       else
             {
             /* Here is a big else condition.
             We must find a process node with the same data or a
             null field */

          current_ptr = entity_ptr->ptr2;
```

```
while (current_ptr->ptr1 ! = NULL)
{

 if (strcmp(current_ptr->data, yytext) = = 0) break;
 current_ptr = current_ptr->ptr1;

}

/*printf("CURRENT PROCESS NAME %s YYTEXT IS %s\n", cur-
rent_ptr->data, yytext);*/

/*If no process matching the name exists, create a new one.
*/

if (strcmp(current_ptr->data, yytext) = = 0)
 { /*printf("YACC: THE CURRENT PROCESS IS ALREADY REGIS-
 TERED\n");*/}
else
 { /*printf("————————THE TWO PROCS ARE DIFFERENT\n");*/
 new_ptr = (struct node *)malloc(sizeof(struct node));
 (void)strcpy(new_ptr->name, "PROCESS");
 (void)strcpy(new_ptr->data, yytext);
 new_ptr->ptr1 = NULL;
 new-ptr->ptr2 = NULL;
 new_ptr->ptr3 = NULL;
 new_ptr->ptr4 = NULL;
 current_ptr->ptr1 = new_ptr;
 current_ptr = new_ptr;
  /*printf("CREATED A NEW PROCESS %s\n", current_ptr-
  >data);*/}

 /* Create the corresponding statement node */

 new_ptr = (struct node *)malloc(sizeof(struct node));
 (void)strcpy(new_ptr->name, "STATEMENT");
 (void)strcpy(new_ptr->data, "UNDEF");
 new_ptr->ptr1 = NULL;
 new_ptr->ptr2 = NULL;
 new_ptr->ptr3 = NULL;
 new_ptr->ptr4 = NULL;
 proc_ptr = current_ptr;

 if (proc_ptr->ptr2 = = NULL)
  {
  current_ptr->ptr2 = new_ptr;
  current_ptr = new_ptr;
  }
 else
  {
  current_ptr = proc_ptr->ptr2;
  while (current_ptr->ptr1 ! = NULL)
  {current_ptr = current_ptr->ptr1;}
  current_ptr->ptr1 = new_ptr;
  current_ptr = new_ptr;
  }

 }
stat_ptr = current_ptr;

/* At this point, current_ptr (stat_ptr) marks the last
statement node (unfilled). I use stat ptr as a precaution
```

```
        against a corrupted current_ptr later in the code. This
        could possibly occur as a result of the extensive use of
        the pointer at different times*/

        }
        | behaviour_header '(' behaviour_parameters')'
        ;

behaviour_parameters:INTEGER
         {/*printf("yacc: behaviour parameter %s\n", yytext);*/
         if (dsl_state = = 1) (void)strcat(scratch, ",");
         }
        | DSL_VAR
         {/*printf("yacc: behaviour parameter %s\n", yytext);*/
         if (dsl_state = = 1) (void)strcat(scratch, ",");
         }
        | behaviour_parameters ',' DSL_VAR
         {/*printf("yacc: %d compund parameter %s\n", dsl_state,
         yytext);*/
         if (dsl_state = = 1) (void)strcat(scratch, ",");
         }
        | behaviour_parameters ',' INTEGER
         {/*printf("yacc: compund parameter %s\n", yytext);*/
         if (dsl_state = = 1) (void)strcat(scratch, ",");
         }
        | behaviour_parameters ',' behaviour_name
         {dsl_state = 0;}
        | behaviour_name
         {dsl_state = 0;}
        ;

behaviour_name:DSL_NAME
         {/*printf("yacc: compund name %s\n", yytext);*/
         dsl_state = 1; (void)strcpy(resource_name,
         yytext);(void)strcpy(scratch, "(");
         }
        | behaviour_name '(' behaviour_parameters ')'
         {dsl_state = 0;}
         ;

behaviour_param:DSL_VAR
        {/*printf("recognized var %s\n", yytext);*/

        }
        | DSL_NAME
        {/*printf("recognized name %s\n", yytext);*/

        }
        | INTEGER
        {/*printf("recognized value %s\n", yytext);*/

        }
        ;

behaviour_statement:{
        scratch[0] = '\0';
        }
        expr
        | behaviour_statement ',' expr
        ;
expr:   action_statement
        {printf("YACC: evaluating an expression...\n");
```

```
/*
new_ptr = (struct node *)malloc(sizeof(struct node));
(void)strcpy(new_ptr->name,"COMMAND");
new_ptr->ptr1 = NULL;
new_ptr->ptr2 = NULL;
new_ptr->ptr3 = NULL;
new_ptr->ptr4 = NULL;

switch(current_command)
  {
  case 'S':

    (void)strcat(resource_name, ": = ");
    if (scratch[strlen(scratch)-1] = = ',')
    scratch[strlen(scratch)-1] = '\0';
    (void)strcat(resource_name, scratch);

    (void)strcpy(new_ptr->data, resource_name);
    (void)strcpy(new_ptr->name, "SET");

    break;
  case 'D':
    (void)strcpy(new_ptr->data, " AFTER ");
    if (scratch[strlen(scratch)-1] = = ',')
    scratch[strlen(scratch)-1] = '\0';
    (void)strcat(new_ptr->data, scratch);
    (void)strcpy(new_ptr->name, "DELAY");
    (void)strcpy(delay_buffer, new_ptr->data);
    break;
  case 'C':
    (void)strcpy(new_ptr->name, "CHECK");
    (void)strcpy(new_ptr->data, resource_name);
    if (scratch[strlen(scratch)-1] = = ',')
    scratch[strlen(scratch)-1] = '\0';
    (void)strcpy(new_ptr->data1, scratch);

    break;
  case 'c':
    (void)strcpy(new_ptr->name, "CREATE");
    (void)strcpy(new_ptr->data, resource_name);
    if (scratch[strlen(scratch)-1] = = ',')
    scratch[strlen(scratch)-1] = '\0';
    (void)strcpy(new_ptr->data1, scratch);
    break;
  case 'P':
    (void)strcpy(new_ptr->data1, "PROBE");
    break;
  case 'W':
    (void)strcpy(new_ptr->name, "WRITE");
    break;
  case 'R':
    (void)strcpy(new_ptr->name, "REMOVE");
    (void)strcpy(new-ptr->data, resource_name);
    if (scratch[strlen(scratch)-1] = = ',')
    scratch[strlen(scratch)-1] = '\0';
    (void)strcpy(new_ptr->data1, scratch);
    break;
  case 'A':
    (void)strcpy(new_ptr->name, "ARITH");
    if (scratch[strlen(scratch)-1] = = ',')
    scratch[strlen(scratch)-1] = '\0';
    (void)strcpy(new_ptr->data, scratch);
```

```
            break;
            }*/
            if (current_command = = 'A')
            {(void)strcpy(new_ptr->name, "ARITH");
            if (scratch[strlen(scratch)-1] = = ',')
            scratch[strlen(scratch)-1] = '\0';
            (void)strcpy(new_ptr->data, scratch);
            }
        printf("YACC:PAST THE CASE STATEMENT");
        /*current_ptr = stat_ptr;
        if (current_ptr->ptr2 = = NULL)
         {
         stat_ptr->ptr2 = new_ptr;
         temp_ptr = new_ptr;
         }
        else
         {
         temp_ptr = current_ptr->ptr2;
         while (temp_ptr->ptr1 ! = NULL)
          {temp_ptr = temp_ptr->ptr1;
          }
         temp_ptr->ptr1 = new_ptr;
         temp_ptr = new_ptr;
         }

        */
        scratch[0] = '\0';
    }
    | comm_statement
    {
       current_ptr = stat_ptr;
       if (current_ptr->ptr2 = = NULL)
         {
         stat_ptr->ptr2 = new_ptr;
         temp_ptr = new_ptr;
         }
       else
     {
     temp_ptr = current_ptr->ptr2;
     while (temp_ptr->ptr1 ! = NULL)
     {temp_ptr = temp_ptr->ptr1;
     }
     temp_ptr->ptr1 = new_ptr;
     temp_ptr = new_ptr;
     }
/*printf("YACC:NODE REPORT: %s %s %s\n", temp_ptr->name, temp_ptr-
>data,temp_ptr->data1);*/

}
| DSL_NAME
{/*printf("recognized another command = %s\n", yytext);*/}
;

action_statement:set_header '(' resource_name ',' variables ')'
        | check_header '(' resource_name ',' variables ')'
        | delay_header '(' time_value ')'
        | probe_header '(' behaviour_param ')'
        | write_header '(' write_body ')'
        | create_header '(' resource_name ',' compound_parameters
        ')'
        | remove_header '(' resource_name ',' compound_parameters
        ')'
```

```
        {/*printf("Recognized a remove command\n");*/}
      | arithmetic
      ;

set_header:SETRES
        {printf("recognized a set command \n");
        current_command = 'S';
         new_ptr = (struct node *)malloc(sizeof(struct node));
         (void)strcpy(new_ptr->name, "SET");
         new_ptr->ptr1 = NULL;
         new_ptr->ptr2 = NULL;
         new_ptr->ptr3 = NULL;
         new_ptr->ptr4 = NULL;

        current_ptr = stat_ptr;
        if (current_ptr->ptr2 = = NULL)
         {
         stat_ptr->ptr2 = new_ptr;

         }
        else
         {
         temp_ptr = current_ptr->ptr2;
         while (temp_ptr->ptr1 ! = NULL)
         {temp_ptr = temp_ptr->ptr1;
         }
        temp_ptr->ptr1 = new_ptr;

         }
        current_ptr = new_ptr;

        }
        ;

create_header:CREATE
        {/*printf("recognized a create command\n");*/
        current_command = 'c';
        new_ptr = (struct node *)malloc(sizeof(struct node));
        (void)strcpy(new_ptr->name, "CREATE");
        new_ptr->ptr1 = NULL;
        new_ptr->ptr2 = NULL;
        new_ptr->ptr3 = NULL;
        new_ptr->ptr4 = NULL;

        current_ptr = stat_ptr;
        if (current_ptr->ptr2 = = NULL)
        {
        stat_ptr->ptr2 = new_ptr;

        }
        else
        {
        temp_ptr = current_ptr->ptr2;
        while (temp_ptr->ptr1 ! = NULL)
        {temp_ptr = temp_ptr->ptr1;
        }
        temp_ptr->ptr1 = new_ptr;

        }
      current_ptr = new_ptr;

      }
      ;
```

```
check_header:CHECKRES
        {printf("recognized a check command \n");
        current_command = 'C';send_state = 0;

        new_ptr = (struct node*)malloc(sizeof(struct node));
        (void)strcpy(new_ptr->name, "CHECK");
        new_ptr->ptr1 = NULL;
        new_ptr->ptr2 = NULL;
        new_ptr->ptr3 = NULL;
        new_ptr->ptr4 = NULL;

        current_ptr = stat_ptr;
        if (current_ptr->ptr2 = = NULL)
        {
        stat_ptr->ptr2 = new_ptr;

        }
        else
        {
        temp_ptr = current_ptr->ptr2;
        while (temp_ptr->ptr1 ! = NULL)
        {temp_ptr = temp_ptr->ptr1;
        }
        temp_ptr->ptr1 = new_ptr;

        }
        current_ptr = new_ptr;
        /* new_ptr is the node for the current command, and we use
        temp_ptr for nodes related to it. */

    }
    ;
remove_header:REMOVE
        {current_command = 'R';
        new_ptr = (struct node *)malloc(sizeof(struct node));
        (void)strcpy(new_ptr->name, "REMOVE");
        new_ptr->ptr1 = NULL;
        new_ptr->ptr2 = NULL;
        new_ptr->ptr3 = NULL;
        new_ptr->ptr4 = NULL;

    current_ptr = stat_ptr;
        if (current_ptr->ptr2 = = NULL)
        {
        stat_ptr->ptr2 = new_ptr;

        }
        else
        {
        temp_ptr = current_ptr->ptr2;
        while (temp_ptr->ptr1 ! = NULL)
        {temp_ptr = temp_ptr->ptr1;
        }
        temp_ptr->ptr1 = new_ptr;

        }
        current_ptr = new_ptr;

        }
        ;
delay_header:DELAY
```

```
        {/*printf("recognized a delay command");*/
        current_command = 'D';
         new_ptr = (struct node *)malloc(sizeof(struct node));
         (void)strcpy(new_ptr->name, "DELAY");
         new_ptr->ptr1 = NULL;
         new_ptr->ptr2 = NULL;
         new_ptr->ptr3 = NULL;
         new_ptr->ptr4 = NULL;

         current_ptr = stat_ptr;
         if (current_ptr->ptr2 = = NULL)
          {
          stat_ptr->ptr2 = new_ptr;

          }
         else
          {
          temp_ptr = current_ptr->ptr2;
         while (temp_ptr->ptr1 ! = NULL)
         {temp_ptr = temp_ptr->ptr1;
         }
         temp_ptr->ptr1 = new_ptr;

         }
      current_ptr = new_ptr;
      }
      ;
time_value:INTEGER
        {/*printf("with Delay parameter %s\n", yytext);*/
        (void)strcpy(scratch, yytext);
        }
        | DSL_VAR
        {/*printf("with Delay variable %s\n", yytext);*/
        (void)strcpy(scratch, yytext);
        }
        ;
probe_header:PROBE
        {printf("recognized a probe command ");
        current_command = 'P';}
        ;
send_header:SEND
      {

      send_state = 0;
        new_ptr = (struct node *)malloc(sizeof(struct node));
        (void)strcpy(new_ptr->name, "COMMAND");
        new_ptr->ptr1 = NULL;
        new_ptr->ptr2 = NULL;
        new_ptr->ptr3 = NULL;
        new_ptr->ptr4 = NULL;

        current-command = 's';
        current_ptr = new_ptr;
        (void)strcpy(new_ptr->name, "SEND");}
        ;
write_header:WRITE
        {
        current_command = 'W';}
        ;

write-body:COMMENTS
      {
```

```
        }
        | DSL_VAR
        ;
resource_name:DSL_NAME
        {
        (void)strcpy(resource_name, yytext);
        (void)strcpy(new_ptr->data, yytext);
        }
        | resource_name '(' compound_parameters ')'
        {
        (void)strcat(resource_name, "(");
        if (scratch[strlen(scratch)-1] = = ',')
        scratch[strlen(scratch)-1] = '\0';
        (void)strcat(resource_name, scratch);
        (void)strcat(resource_name,")" );
        (void)strcpy(scratch, " ");
        }
        ;
compound_parameters:INTEGER
        {/*printf("yacc: compund parameter %s\n", yytext);*/
        (void)strcat(scratch, yytext);
        (void)strcat(scratch, ",");

        /* the first maloc for temp_ptr is to place a node for the
        type tracking */
        temp_ptr = (struct node *)malloc(sizeof(struct node));
        (void)strcpy(temp_ptr->name, yytext);
        (void)strcpy(temp_ptr->data, "INT");
        temp_ptr->ptr1 = NULL;
        temp_ptr->ptr2 = NULL;
        temp_ptr->ptr3 = NULL;
        temp_ptr->ptr4 = NULL;

        if (proc_ptr->ptr4 = = NULL)
        {
        proc_ptr->ptr4 = temp_ptr;
        }
        else
        {
        another_ptr = proc_ptr->ptr4;
        while (another_ptr->ptr1 ! = NULL)
        {
        if (strcmp(another_ptr->name, yytext) = = 0) break;
        another_ptr = another_ptr->ptr1;
        }
        if (strcmp(another_ptr->name, yytext) = = 0)
        {
        free(temp_ptr);
        }
        else
        {another_ptr->ptr1 = temp_ptr;}
        }

        /* Now, we malloc again to store the various parameters */

        /*if (send_state = = 1)*/

        if (send_state = = 2)
        {(void) strcat(temp_ptr->data, yytext);
        (void) strcat(temp_ptr->data, ",");
        }
        else
```

```
{

temp_ptr = (struct node *)malloc(sizeof(struct node));
(void)strcpy(temp_ptr->name, yytext);
(void)strcpy(temp_ptr->data, "INTEGER");
temp_ptr->ptr1 = NULL;
temp_ptr->ptr2 = NULL;
temp_ptr->ptr3 = NULL;
temp_ptr->ptr4 = NULL;

if (new_ptr->ptr2 = = NULL)
{/*printf("NEW PTR NAME IS %s\n", new_ptr->name);*/
new_ptr->ptr2 = temp_ptr;

}
else
{/*printf("NEW PTR NAME IS %s----%s\n", new_ptr->name,
new_ptr->ptr2);*/
another_ptr = new_ptr->ptr2;
while (another_ptr->ptr1 ! = NULL)
{another_ptr = another_ptr->ptr1;
}
another_ptr->ptr1 = temp_ptr;
}

}
}
| DSL_VAR
{printf("yacc: compound parameter %s SRATCH %s\n", yytext,
scratch);

(void)strcat(scratch, yytext);
(void)strcat(scratch, ",");

temp_ptr = (struct node *)malloc(sizeof(struct node));
(void)strcpy(temp_ptr->name, yytext);
(void)strcpy(temp_ptr->data, "VAR");
temp_ptr->ptr1 = NULL;
temp_ptr->ptr2 = NULL;
temp_ptr->ptr3 = NULL;
temp_ptr->ptr4 = NULL;

if (proc_ptr->ptr4 = = NULL)
{
proc_ptr->ptr4 = temp_ptr;
}
else
{
another_ptr = proc_ptr->ptr4;
while (another_ptr->ptr1 ! = NULL)
{
if (strcmp(another_ptr->name, yytext) = = 0) break;
another_ptr = another_ptr->ptr1;
}
if (strcmp(another_ptr->name, yytext) = = 0)
{
free(temp_ptr);
}
else
{another_ptr->ptr1 = temp_ptr;}
}
```

```
       switch(send_state)
       {
       case 2:printf("IN CASE2\n");
       (void)strcat(temp_ptr->data, yytext);
       (void)strcat(temp_ptr->data, ",");
       break;

       case 1:printf("IN CASE 1\n");
       temp_ptr = (struct node *)malloc(sizeof(struct node));
       (void)strcpy(temp_ptr->name, yytext);
       (void)strcpy(temp_ptr->data, "VAR");
       (void)strcpy(temp_ptr->data1, resource_name);
       temp_ptr->ptr1 = NULL;
       temp_ptr->ptr2 = NULL;
       temp_ptr->ptr3 = NULL;
       temp_ptr->ptr4 = NULL;

       if (new_ptr->ptr3 = = NULL)
       {
       new_ptr->ptr3 = temp_ptr;

       current_ptr = temp_ptr;

       }
       else
       {
       /*another_ptr = new_ptr->ptr3;
       while (another_ptr->ptr1 ! = NULL)
       {another_ptr = another_ptr->ptr1;
       }
       another_ptr->ptr1 = temp_ptr;*/
       current_ptr->ptr1 = temp_ptr;
       current_ptr = temp_ptr;

       }
       break;
case 0:printf("IN 1st\n");
       temp_ptr = (struct node *)malloc(sizeof(struct node));
       (void)strcpy(temp_ptr->name, yytext);
       (void)strcpy(temp_ptr->data, "VAR");
       (void)strcpy(temp_ptr->data1, resource_name);
       temp_ptr->ptr1 = NULL;
       temp_ptr->ptr2 = NULL;
       temp_ptr->ptr3 = NULL;
       temp_ptr->ptr4 = NULL;

       if (new_ptr->ptr2 = = NULL)
       {
       new_ptr->ptr2 = temp_ptr;
       current_ptr = temp_ptr;
       }
       else
       {
       /*another_ptr = new_ptr->ptr2;
       while (another_ptr->ptr1 ! = NULL)
       {another_ptr = another_ptr->ptr1;
       }
       another_ptr->ptr1 = temp_ptr;*/
       current_ptr->ptr1 = temp_ptr;
       current_ptr = temp_ptr;
       }
       break;
```

```
    }
    /* ABOVE: case 2 is a message in a send, 1 is a port in a
    send, 0 is a compound parameter in general */
    }

| compound_parameters ´,´ DSL_VAR
    {/*printf(¨yacc: compund parameter %s\n¨, yytext);*/
    (void)strcat(scratch, yytext);
    (void)strcat(scratch, ¨,¨);
    temp_ptr = (struct node *)malloc(sizeof(struct node));
    (void)strcpy(temp_ptr->name, yytext);
    (void)strcpy(temp_ptr->data, ¨VAR¨);
    temp_ptr->ptr1 = NULL;
    temp_ptr->ptr2 = NULL;
    temp_ptr->ptr3 = NULL;
    temp_ptr->ptr4 = NULL;

    if (proc_ptr->ptr4 = = NULL)
    {
    proc_ptr->ptr4 = temp_ptr;
    }
    else
    {
    another_ptr = proc_ptr->ptr4;
    while (another_ptr->ptr1 ! = NULL)
    {
    if (strcmp(another_ptr->name, yytext) = = 0) break;
    another_ptr = another_ptr->ptr1;
    }
    if (strcmp(another_ptr->name, yytext) = = 0)
    {
    free(temp_ptr);
    }
    else
    {another_ptr->ptr1 = temp_ptr;}
    }

    switch(send_state)
    {
    case 2:
    (void)strcat(temp_ptr->data, yytext);
    (void)strcat(temp_ptr->data, ¨,¨);
    break;

    case 1:
    temp_ptr = (struct node *)malloc(sizeof(struct node));
    (void)strcpy(temp_ptr->name, yytext);
    (void)strcpy(temp_ptr->data, ¨VAR¨);
    (void)strcpy(temp_ptr->data1, resource_name);
    temp_ptr->ptr1 = NULL;
    temp_ptr->ptr2 = NULL;
    temp_ptr->ptr3 = NULL;
    temp_ptr->ptr4 = NULL;

    if (new_ptr->ptr3 = = NULL)
    {
    new_ptr->ptr3 = temp_ptr;

    current_ptr = temp_ptr;

    }
    else
    {
```

```
        /*another_ptr = new_ptr->ptr3;
        while (another_ptr->ptr1 ! = NULL)
        {another_ptr = another_ptr->ptr1;
        }
        another_ptr->ptr1 = temp_ptr;*/
        current_ptr->ptr1 = temp_ptr;
        current_ptr = temp_ptr;

        }
        break;
        case 0:printf("IN 2nd\n");
        temp_ptr = (struct node *)malloc(sizeof(struct node));
        (void)strcpy(temp_ptr->name, yytext);
        (void)strcpy(temp_ptr->data, "VAR");
        (void)strcpy(temp_ptr->data1, resource_name);
        temp_ptr->ptr1 = NULL;
        temp_ptr->ptr2 = NULL;
        temp_ptr->ptr3 = NULL;
        temp_ptr->ptr4 = NULL;

        if (new_ptr->ptr2 = = NULL)
        {
        new_ptr->ptr2 = temp_ptr;
        current_ptr = temp_ptr;

        }
        else
        {
        /*another_ptr = new_ptr->ptr2;
        while (another_ptr->ptr1 ! = NULL)
        {another_ptr = another_ptr->ptr1;
        }
        another_ptr->ptr1 = temp_ptr;*/
        current_ptr->ptr1 = temp_ptr;
        current_ptr = temp_ptr;

        }
        break;
        }

        }
        | compound_parameters ',' INTEGER
        {/*printf("yacc: compund parameter %s\n", yytext);*/
        (void)strcat(scratch, ", (INT) ");
        (void)strcat(scratch, yytext);

        if (send_state = = 2)
        {(void)strcat(temp_ptr->data, yytext);
        (void)strcat(temp_ptr->data, ",");
        }

        }
        | compound_parameters ',' compound_name
        {
        /*scratch[strlen(scratch)-2] = '\0';*/
        }
        | compound_name
        ;
compound_name:DSL_NAME
        {  /*printf("yacc: compund name %s\n", yytext);*/
        /* (void)strcat(scratch, "(UNDEF) ");*/
        (void)strcat(scratch, yytext);(void)strcat(scratch, "{");
```

```
if (send_state = = 1)

{ send_state = 2;
/*send_state = 2->that this dslname is treated as an array;
hence it is maintained in one field (data). Scratch is simply
catted onto it. The pointer temp_ptr will be the correct
operating pointer at this time */

temp_ptr = (struct node *)malloc(sizeof(struct node));
(void)strcpy(temp_ptr->name, yytext);
(void)strcpy(temp_ptr->data, ¨ ¨);
scratch[0] = ´\0´;
temp_ptr->ptr1 = NULL;
temp_ptr->ptr2 = NULL;
temp_ptr->ptr3 = NULL;
temp_ptr->ptr4 = NULL;

if (new_ptr->ptr2 = = NULL)
{
new_ptr->ptr2 = temp_ptr;

}
else
{
another_ptr = new_ptr->ptr2;
while (another_ptr->ptr1 ! = NULL)
{another_ptr = another_ptr->ptr1;
}
another_ptr->ptr1 = temp_ptr;
}

}
/* the condition below is true only if this is the first dsl
name encountered in the command
if (send_state = = 0)
{
strcpy(new_ptr->data, yytext);strcpy(new_ptr->data1,
¨UNDEF¨);send_state = 1;
}*/

}
| compound_name ´(´ compound_parameters ´)´
{/* reset the state back to 1, and clean the last 2 charac-
ters from data */
scratch[strlen(scratch)-1] = ´\0´;
strcat(scratch, ¨},¨);
if (send_state = = 2)
{
send_state = 1;
temp_ptr->data[strlen(temp_ptr->data)-1] = ´\0´;
}
}
;
variables: {
scratch[0] = ´\0´;
}
INTEGER
{

(void)strcat(scratch, yytext);
(void)strcat(scratch, ¨,¨);
```

```
temp_ptr = (struct node *)malloc(sizeof(struct node));
(void)strcpy(temp_ptr->name, yytext);
(void)strcpy(temp_ptr->data, "INT");
temp_ptr->ptr1 = NULL;
temp_ptr->ptr2 = NULL;
temp_ptr->ptr3 = NULL;
temp_ptr->ptr4 = NULL;

if (proc_ptr->ptr4 = = NULL)
{
proc_ptr->ptr4 = temp_ptr;
}
else
{
another_ptr = proc_ptr->ptr4;
while (another_ptr->ptr1 ! = NULL)
{
if (strcmp(another_ptr->name, yytext) = = 0) break;
another_ptr = another_ptr->ptr1;
}
if (strcmp(another_ptr->name, yytext) = = 0)
{
free(temp_ptr);
}
else
{another_ptr->ptr1 = temp_ptr;}
}

temp_ptr = (struct node*)malloc(sizeof(struct node));
(void)strcpy(temp_ptr->name, yytext);
(void)strcpy(temp_ptr->data, "INT");
temp_ptr->ptr1 = NULL;
temp_ptr->ptr2 = NULL;
temp_ptr->ptr3 = NULL;
temp_ptr->ptr4 = NULL;

if (new_ptr->ptr3 = NULL)
{ new_ptr->ptr3 = temp_ptr;}
else
{
another_ptr = new_ptr->ptr3;
while (another_ptr->ptr1 ! = NULL)
{another_ptr = another_ptr->ptr1;
}
another_ptr->ptr1 = temp_ptr;
}
}
| DSL_VAR
{

(void)strcat(scratch, yytext);
(void)strcat(scratch, ",");

temp_ptr = (struct node *)malloc(sizeof(struct node));
(void)strcpy(temp_ptr->name, yytext);
(void)strcpy(temp_ptr->data, "VAR");
temp_ptr->ptr1 = NULL;
temp_ptr->ptr2 = NULL;
temp_ptr->ptr3 = NULL;
temp_ptr->ptr4 = NULL;

if (proc_ptr->ptr4 = = NULL)
```

```
{
proc_ptr_>ptr4 = temp_ptr;
}
else
{
another_ptr = proc_ptr->ptr4;
while (another_ptr->ptr1 ! = NULL)
{
if (strcmp(another_ptr->name, yytext) = = 0) break;
another_ptr = another_ptr->ptr1;
}
if (strcmp(another_ptr->name, yytext) = = 0)
{
free(temp_ptr);
}
else
{another_ptr->ptr1 = temp_ptr;}
}
temp_ptr = (struct node *)malloc(sizeof(struct node));
(void)strcpy(temp_ptr->name, yytext);
(void)strcpy(temp_ptr->data, "VAR");
temp_ptr->ptr1 = NULL;
temp_ptr->ptr2 = NULL;
temp_ptr->ptr3 = NULL;
temp_ptr->ptr4 = NULL;

if (new_ptr->ptr3 = = NULL)
{ new_ptr->ptr3 = temp_ptr;}
else
{
another_ptr = new_ptr->ptr3;
while (another_ptr->ptr1 ! = NULL)
{another_ptr = another_ptr->ptr1;
}
another_ptr->ptr1 = temp_ptr;
}

}
| DSL_NAME
{
/*(void)strcat(scratch, " (UNDEF) ");*/
(void)strcat(scratch, yytext);

temp_ptr = (struct node *)malloc(sizeof(struct node));
(void)strcpy(temp_ptr->name, yytext);
(void)strcpy(temp_ptr->data, "UNDEF");
temp_ptr->ptr1 = NULL;
temp_ptr->ptr2 = NULL;
temp_ptr->ptr3 = NULL;
temp_ptr->ptr4 = NULL;

if (proc_ptr->ptr4 = = NULL)
{
proc_ptr->ptr4 = temp_ptr;
}
else
{
another_ptr = proc_ptr->ptr4;
while (another_ptr->ptr1 ! = NULL)
{
if (strcmp(another_ptr->name, yytext) = = 0) break;
another_ptr = another_ptr->ptr1;
```

```
      }
      if (strcmp(another_ptr->name, yytext) = = 0)
      {
      free(temp_ptr);
      }
      else
      {another_ptr->ptr1 = temp_ptr;}
      }

      temp_ptr = (struct node *)malloc(sizeof(struct node));
      (void)strcpy(temp_ptr->name, yytext);
      (void)strcpy(temp_ptr->data, ¨UNDEF¨);
      temp_ptr->ptr1 = NULL;
      temp_ptr->ptr2 = NULL;
      temp_ptr->ptr3 = NULL;
      temp_ptr->ptr4 = NULL;

      if (new_ptr->ptr3 = = NULL)
      { new_ptr->ptr3 = temp_ptr;}
      else
      {
      another_ptr = new_ptr->ptr3;
      while (another_ptr->ptr1 ! = NULL)
      {another_ptr = another_ptr->ptr1;
      }
      another_ptr->ptr1 = temp_ptr;
      }

      }
      | ´(´ compound_parameters ´)´
      ;

arithmetic:DSL_VAR OPERATOR expression
      {/*printf(¨yacc recognized an arithmetic expression \n¨);*/
      current_command = ´A´;

      new_ptr = (struct node *)malloc(sizeof(struct node));
      (void)strcpy(new_ptr->name, ¨ARITH¨);
      new_ptr->ptr1 = NULL;
      new_ptr->ptr2 = NULL;
      new_ptr->ptr3 = NULL;
      new_ptr->ptr4 = NULL;

      current_ptr = stat_ptr;
      if (current_ptr->ptr2 = = NULL)
      {
      stat_ptr->ptr2 = new_ptr;

      }
      else
      {
      temp_ptr = current_ptr->ptr2;
      while (temp_ptr->ptr1 ! = NULL)
      {temp_ptr = temp_ptr->ptr1;
      }
      temp_ptr->ptr1 = new_ptr;

      }
      }
      ;
expression:DSL_VAR
      | INTEGER
```

```
        | expression OPERATOR
        | expression DSL_VAR
        | expression INTEGER
        | '(' expression ')'
        ;
comm_statement: send_statement
        ;

send_statement:{scratch[0] = '\0';}
        send_header '(' send_parameters ')'
        {/*printf("Send command matched completely \n");*/}
        ;

send_parameters:dest_name ',' port_name ',' message
        {

        }
        | dest_name
        { /* Here destname refers to a message */
        (void)strcpy(new_ptr->data1, resource_name);
        (void)strcpy(new_ptr->data, "INTERNAL");
        }
        ;

port_name:
        DSL_NAME
        {/*printf("port of send = %s\n", yytext);*/
        (void)strcpy(new_ptr->data, yytext); send_state = 1;}
        | DSL_NAME '(' compound_parameters ')'
        {printf("YACC: port of send = %s\n", resource_name);
        if (strlen(scratch)>3)
        { /*printf("YACC: port of send = %s\n", resource_name);*/
        (void)strcpy(new_ptr->data, rsource_name);
        (void)strcat(new_ptr->data, "(");
        scratch[strlen(scratch)-1] = '\0';
        (void)strcat(new_ptr->data, scratch);
        (void)strcat(new_ptr->data, ")");
        /*printf("YACC: PORT = = %s\n", new_ptr->data);*/
        scratch[0] = '\0';send_state = 1;
        }

        }
        | DSL_VAR
        {(void)strcpy(new_ptr->data, yytext);send_state = 1;}
        ;

dest_name:DSL_NAME
        {  /*printf("destination module of send = %s\n", yytext);*/
        (void)strcpy(resource_name, yytext);scratch[0] = '\0';
        }
        | dest_name '(' compound_parameters ')'
        {  strcat(resource_name,"(");
        strcat(resource_name, scratch);
        strcat(resource_name,")");scratch[0] = '\0';
        }
        | '[' group_dest ']'
        | '-'
        {/*printf("lex: no destination given\n");*/}
        ;

group_dest:DSL_NAME
        | group_dest ',' DSL_NAME
```

```
        ;

message: DSL_NAME
        {
        /*(void)strcpy(scratch, yytext);*/
        (void)strcpy(new_ptr->data1, yytext);

        }

        | message ´(´ compound_parameters ´)´
        {
        /*(void)strcpy(new_ptr->data1, resource_name); */

        scratch[0] = ´\0´; send_state = 0;
        }

        ;

ho-definition:HOMODULE ´(´ module_name ´,´ ´[´ sub_modules ENDER
        | HOMODULE ´(´ module_name ´,´ ´[´ ENDER
        ;

sub-modules:module_name
        | sub_modules ´,´ module_name
        ;

%%
#include ¨lex.yy.c¨
main()
{
yyparse();

        printf(¨ \n\n\n\nLISTING PROCESSES\n¨);
        entity_ptr = head_ptr;
        while(entity_ptr ! = 0)
        {
        printf(¨\nFOR MODULE/ENTITY %s\n\n¨, entity_ptr->data);
        proc_ptr = entity_ptr->ptr2;
        while(proc_ptr ! = NULL)
        {printf(¨  LISTING %s %s\n¨, proc_ptr->name, proc_ptr->data);
        proc_ptr = proc_ptr->ptr1;}
        entity_ptr = entity_ptr->ptr1;
        }
}
```

Appendix

D

Lex Listing for Parser

The following is the lex listing for the DSL-VHDL parser.

```
%e 700
%p 3100
%n 500
%a 2600
%o 3900
%k 1100
D      [0-9]
B      [\t\n]
W       {B}+
OW       {B}*
N      [a-z][a-z0-9_]*
V                   [A-Z][a-zA-Z0-9_]*
UN                  [_][A-Z0-9]*
INT    [0-9][0-9]*
REAL               ({D}+"."{D}*|{D}*"."{D}+)
OPS                (is|\+|V|\-|\>|\<|\*)
X      [^\t\n]+
OPEN   [\(]
LIT   [a-zA-Z0-9_\-\*]*
STRING                   "'"{LIT}"'"
NUMBER                   ({INT}|{REAL})
DSLO  (other|variables|delay)
DSLH              (module|ho_module|path|isa)
%S M_HEADER M_NAME B_HEADER B_PAR B_BODY M_VAR
%S PARAMS PARAMS1 PARAMS2 PARAMS3 B_ARITH
%S HO_HEADER HO_NAME HO_MODULES HO_SUB_NAME
%%
^ho_module       {BEGIN HO_HEADER;
                   printf("detected an HO-module \n");
                   return (HOMODULE);
                   }
<HO_HEADER>{OPEN}return yytext[0];
<HO_HEADER>{N}         {BEGIN HO_NAME;
                   printf("lex: detected ho name %s\n", yytext);
                   return (DSL_NAME);
                   }
<HO_NAME>\(     return yytext[0];
<HO_NAME>{V}    return (DSL_VAR);
<HO_NAME>{INT}  return (INTEGER);
```

```
<HO_NAME>\)      return yytext[0];
<HO_NAME>\,      return yytext[0];
<HO_NAME>\[      {BEGIN HO_MODULES;
                   printf("lex: Detected sqb-expecting a list\n");
                   return yytext[0];
                       }
<HO_NAME>\]\)\.        {printf("lex:matched end of ho-module \n");
                          return (ENDER);
                          }
<HO_MODULES>{N}        {printf("lex: A submodule for this ho %s\n",
                          yytext);
                          return (DSL_NAME);
                          }
<HO_MODULES>\(         {BEGIN HO_SUB_NAME;
                          return yytext[0];
                          }
<HO_MODULES>\,         return yytext[0];
<HO_SUB_NAME>{N}return (DSL_NAME);
<HO_SUB_NAME>{V}return (DSL_VAR);
<HO_SUB_NAME>{INT}{printf("lex: subparameter is %s\n", yytext);
                          return (INTEGER);
                          }
<HO_SUB_NAME>\(        return yytext[0];
<HO_SUB_NAME>\)        {BEGIN HO_MODULES;
                          printf("lex: Another submodule bites the
                          dust…\n");
                          return yytext[0];
                              }
<HO_SUB_NAME>\,        return yytext[0];
<HO_MODULES>\]\)\.{printf("———lex: completed scan of ho-
                          module\n");
                          return (ENDER);
                          }
^module                 {BEGIN M_HEADER;
                          printf("lex:detected a module \n");
                          return (MODULE);
                          }
<M_HEADER>{OPEN}{printf("lex:about to return a round brack \n");
                          return yytext[0];
                          }
<M_HEADER>{N}          {BEGIN M_NAME;
                          printf("lex:detected dslname %s\n",
                          yytext);
                          return (DSL_NAME);
                          }
<M_NAME>\(             {BEGIN M_VAR;
                          (void)strcpy(scratch, "VHDL ARCH PARAME-
                          TERS ");
                          printf("lex:detected open brack %s\n",
                          yytext);
                          return yytext[0];
                          }
<M_VAR>{V}             {printf("lex: variable detected for name
                       %s\n", yytext);
                          strcat(scratch, "VAR ");
                          strcat(scratch, yytext);
                          strcat(scratch, "; ");
                          return (DSL_VAR);
                          }
<M_VAR>{INT} {strcat(scratch, "INT ");
                          strcat(scratch, yytext);
                          strcat(scratch, "; ");
```

```
                           return (INTEGER);
                           }
<M_VAR>\,               return yytext[0];
<M_VAR>\)               {BEGIN M_NAME;
                           (void)strcpy(entity_ptr->data1, scratch);
                           /* write parameters to struct */

                           return yytext[0];
                           }
<M_NAME>\,              return yytext[0];
<M_NAME>\[              {BEGIN B_HEADER;
                           printf("lex:detected a square bracket
                           \n");
                           return yytext[0];
                           }
<B_HEADER>\(            return yytext[0];
<B_HEADER>{N}           {BEGIN B_PAR;count = 0;/* begin behavior
                        decompose */
                           printf("lex:detected a dsl name \n");
                           return (DSL_NAME);
                           }
<B_HEADER>\,            return yytext[0];
<B_PAR>\:\-             {BEGIN B_BODY;
                           printf("lex:separator is %s\n", yytext);
                           return (GOAL);
                           }
<B_PAR>\(               return yytext[0];/*searching for parameters
                        of behavior */
<B_PAR>{N}              {printf("lex:returning a dsl name %s\n",
                        yytext);
                           return (DSL_NAME);
                           }
<B_PAR>{INT}            {
                           if (dsl_state ! = 1)
                           {count = count + 1;
                           /* first record the integer as a condition
                           */
                           new_ptr = (struct node
                           *)malloc(sizeof(struct node));
                             (void)strcpy(new_ptr->name, "CONDI-
                             TION");
                             (void)strcpy(new_ptr->data, yytext);
                             (void)strcpy(new_ptr->data1, "INTEGER");
                             new_ptr->ptr1 = NULL;
                             new_ptr->ptr2 = NULL;
                             new_ptr->ptr3 = NULL;
                             new_ptr->ptr4 = NULL;

                             if (stat_ptr->ptr3 ! = NULL)
                             {
                             temp_ptr = stat_ptr->ptr3;

                              while (temp_ptr->ptr1 ! = NULL)
                              {temp_ptr = temp_ptr->ptr1;}

                             temp_ptr->ptr1 = new_ptr;

                             }
                             else
                             {
                             stat_ptr->ptr3 = new_ptr;
                             }
```

```
                        /*Now check to see if an undefined or
                        nonexistent signal exists at position
                        count */

                          new_ptr = (struct node
                          *)malloc(sizeof(struct node));
                          (void)strcpy(new_ptr->name, ¨SIGNAL¨);
                          (void)strcpy(new_ptr->data, ¨UNDEF¨);
                          (void)strcpy(new_ptr->data1, ¨INTEGER¨);
                          new_ptr->ptr1 = NULL;
                          new_ptr->ptr2 = NULL;
                          new_ptr->ptr3 = NULL;
                          new_ptr->ptr4 = NULL;
                        if (proc_ptr->ptr3 = = NULL)
                           {proc_ptr->ptr3 = new_ptr;}
                        else
                        {
                        temp_ptr = proc_ptr->ptr3;
                        icount = count-1;
                          while (icount>0)
                           {if (temp_ptr->ptr1 = = NULL) break;
                           temp_ptr = temp_ptr->ptr1;
                           icount = icount-1;
                        }
                        /* We are now pointing to the correct
                        position signal */
                          if (icount = = 0)
                          {/* we are on a duplicate reference sig-
                          nal */
                          if (strcmp(temp_ptr->data1, ¨UNDEF¨) = =
                          0)
                           {(void)strcpy(temp_ptr->data1, ¨INTE-
                           GER¨);
                           free(new_ptr);}
                          }
                          else
                           {/* A signal node does not exist for
                           this parameter position. Create an
                           undef one */
                           temp_ptr->ptr1 = new_ptr;
                        }
                        }
                        }
                        else {(void)strcat(scratch, yytext);}

                        return (INTEGER);
                        }
<B_PAR>>{V}         {printf(¨lex:returning a dsl variable %s\n¨,
                    yytext);
                        if (dsl_state ! = 1)
                        {count = count + 1;

                          new_ptr = (struct node
                          *)malloc(sizeof(struct node));
                          (void)strcpy(new_ptr->name, ¨SIGNAL¨);
                          (void)strcpy(new_ptr->data1, ¨UNDEF¨);
                          (void)strcpy(new_ptr->data, yytext);
                          new_ptr->ptr1 = NULL;
                          new_ptr->ptr2 = NULL;
                          new_ptr->ptr3 = NULL;
                          new_ptr->ptr4 = NULL;

                        if (proc_ptr->ptr3 ! = NULL)
```

```
{
temp_ptr = proc_ptr->ptr3;

  while (temp_ptr->ptr1 ! = NULL)
  {if (strcmp(temp_ptr->data, yytext) = =
  0) break;
  temp_ptr = temp_ptr->ptr1;
  }
  if (strcmp(temp_ptr->data, yytext) = =
  0) {printf("LEX: Duplicate variable\n");
  free(new_ptr);}
  else {temp_ptr->ptr1 = new_ptr;}
}
else
{
proc_ptr->ptr3 = new_ptr;
}

/* Now check to see if an undefined or
nonexistent signal exists at position
count */

new_ptr = (struct node
*)malloc(sizeof(struct node));
  (void)strcpy(new_ptr->name, "CONDI-
  TION");
  (void)strcpy(new_ptr->data, "UNDEF");
  (void)strcpy(new_ptr->data1, "UNDEF");
  new_ptr->ptr1 = NULL;
  new_ptr->ptr2 = NULL;
  new_ptr->ptr3 = NULL;
  new_ptr->ptr4 = NULL;
if (stat_ptr->ptr3 = = NULL)
  {stat_ptr->ptr3 = new_ptr;}
else
{
temp_ptr = stat_ptr->ptr3;
icount = count - 1;
  while (icount>0)
  {  if (temp_ptr->ptr1 = = NULL) break;
    temp_ptr = temp_ptr->ptr1;
    icount = icount - 1;
  }
  /* We are now pointing to the correct
  position condition */
if (icount = = 0)
  {/* we are on a duplicate reference sig-
  nal */
  free(new_ptr);
  }
else
  {/* A signal node does not exist for
  this parameter position. Create an undef
  one */
  temp_ptr->ptr1 = new_ptr;
  }
}
}
else {(void)strcat(scratch, yytext);}

printf("lex: PRINTING A NEW SIGNAL NODE:
%s %s\n", new_ptr->data1,
new_ptr->data);
```

```
                              return (DSL_VAR);
                              }
<B_PAR>\,                 return yytext[0];
<B_PAR>\)                 {
                              if (dsl_state = = 1)
                              {
                              {count = count + 1;
                              /* first record the integer as a condition
                              */
                              new_ptr = (struct node *)malloc(sizeof
                              (struct node));
                                (void)strcpy(new_ptr-> name, ¨CONDI-
                                TION¨);
                                if (strlen(scratch)>4)
                                {(void)strcat(resource_name, scratch);
                                 resource_name[strlen(resource_name)-1]
                                 = ´\0´;
                                 (void)strcat(resource_name,¨)¨);}
                                (void)strcpy(new_ptr->data,
                                resource_name);
                                (void)strcpy(new_ptr->data1, ¨DSL¨);
                                new_ptr->ptr1 = NULL;
                                new_ptr->ptr2 = NULL;
                                new_ptr->ptr3 = NULL;
                                new_ptr->ptr4 = NULL;

                                if (stat_ptr->ptr3 ! = NULL)
                                {
                                temp_ptr = stat_ptr->ptr3;

                                 while (temp_ptr->ptr1 ! = NULL)
                                 {temp_ptr = temp_ptr->ptr1;}

                                temp_ptr->ptr1 = new_ptr;

                                }
                                else
                                {
                                stat_ptr->ptr3 = new_ptr;
                                }

                              /* Now check to see if an undefined or
                              nonexistent signal exists at position
                              count */

                                new_ptr = (struct node
                                *)malloc(sizeof(struct node));
                                (void)strcpy(new_ptr->name, ¨SIGNAL¨);
                                (void)strcpy(new_ptr->data, ¨UNDEF¨);
                                (void)strcpy(new_ptr->data1, ¨DSL¨);
                                new_ptr->ptr1 = NULL;
                                new_ptr->ptr2 = NULL;
                                new_ptr->ptr3 = NULL;
                                new_ptr->ptr4 = NULL;
                              if (proc_ptr->ptr3 = = NULL)
                                {proc_ptr->ptr3 = new_ptr;}
                              else
                              {
                              temp_ptr = proc_ptr->ptr3;
                              icount = count - 1;
                                while (icount>0)
                                {  if (temp_ptr->ptr1 = = NULL) break;
                                 temp_ptr = temp_ptr->ptr1;
```

```
                                icount = icount - 1;
                              }
                              /* We are now pointing to the correct
                              position signal */
                            if(icount == 0)
                              {/* we are on a duplicate reference sig-
                              nal */
                              if (strcmp(temp_ptr->data1, "UNDEF") ==
                              0)
                                {(void)strcpy(temp_ptr->data1, "DSL");
                                free(new_ptr);}
                              }
                            else
                              {/* A signal node does not exist for
                              this parameter position. Create an undef
                              one */
                              temp_ptr->ptr1 = new_ptr;
                              }
                            }
                            }

                            dsl_state = 0;
                            }
                            return yytext[0];
                            }

<B_BODY>\(               {BEGIN PARAMS;
                            return yytext[0];
                            }
<B_BODY>{V}              {BEGIN B_ARITH;
                              /* flush out scratch buffer and place
                              operand */
                              (void)strcpy(scratch, yytext);
                            return (DSL_VAR);
                            }
<B_BODY>send             {printf("LEX: DETECTED A SEND\n");return
                         (SEND);}

<B_BODY>set_res          return (SETRES);

<B_BODY>check_resreturn (CHECKRES);

<B_BODY>delay            return (DELAY);
<B_BODY>probe            return (PROBE);
<B_BODY>write            return (WRITE);
<B_BODY>create           return (CREATE);
<B_BODY>remove           return (REMOVE);
<B_BODY>print            return (WRITE);
<B_BODY>printl           return (WRITE);
<B_BODY>\,               return yytext[0];
<B_BODY>\)               {BEGIN B_HEADER;
                            return yytext[0];
                            }
<B_HEADER>\]\)\.{print("----lex: completed scan of module \n");
                            return (ENDER);
                            }
<B_ARITH>{OPS}           {printf("lex. passing an operator %s\n",
                         yytext);
                               if (strcmp(yytext, "is") == 0)
                                (void)strcat(scratch, " := ");
                               else {
                                (void)strcat(scratch, " ");
                                (void)strcat(scratch, yytext);
```

```
                                (void)strcat(scratch, " ");
                                }
                             return (OPERATOR);
                             }
<B_ARITH>{V}            {
                             (void)strcat(scratch, yytext);
                             return (DSL_VAR);
                             }
<B_ARITH>{N}            {(void)strcat(scratch, yytext);
                             return (DSL_NAME);
                             }
<B_ARITH>{INT}          {(void)strcat(scratch, yytext);
                             return (INTEGER);
                             }
<B_ARITH>\,             {BEGIN B_BODY;
                             printf("lex:matched an arithmetic expres-
                             sion \n");
                             return yytext[0];
                             }
<PARAMS>\)              {BEGIN B_BODY;
                             return yytext[0];
                             }
<PARAMS>\(              {BEGIN PARAMS1;
                             return yytext[0];
                             }
<PARAMS1>\)             {BEGIN PARAMS;
                             return yytext[0];
                             }
<PARAMS1>\(             {BEGIN PARAMS2;
                             return yytext[0];
                             }
<PARAMS2>\)             {BEGIN PARAMS1;
                             return yytext[0];
                             }
<PARAMS2>\(             {BEGIN PARAMS3;
                             return yytext[0];
                             }
<PARAMS3>\)             {BEGIN PARAMS2;
                             return yytext[0];
                             }
<PARAMS,PARAMS1,PARAMS2,PARAMS3>{N}{printf("lex:matched a name
%s\n", yytext);
                             if (current_command = = 's')
                             {(void)strcpy(resource_name, yytext);
                             printf("LEX: UPDATED SEND PORT NAME %s\n",
                             resource_name);
                             }
                             return (DSL_NAME);
                             }
<PARAMS,PARAMS1,PARAMS2,PARAMS3>\-return yytext[0];
<PARAMS,PARAMS1,PARAMS2,PARAMS3>{V}{printf("lex:MATCHED A DSL VAR
%s\n", yytext);

                             return (DSL_VAR);
                             }
<PARAMS,PARAMS1,PARAMS2,PARAMS3>{INT}return (INTEGER);

<PARAMS,PARAMS1,PARAMS2,PARAMS3>{OPS}return (OPERATOR);
<PARAMS,PARAMS1,PARAMS2,PARAMS3>{STRING}return (COMMENTS);
<PARAMS,PARAMS1,PARAMS2,PARAMS3>\,return yytext[0];
[\t\n]+                 ;
```

Bibliography

ATM Forum, *User-Network Interface Specification (Version 3.0)*, Prentice-Hall, Englewood Cliffs, N.J., 1993.

Bae, Jaime Jungok, and T. Suda, "Survey of Traffic Control Schemes and Protocols in ATM Networks," *Proceedings of the IEEE,* 79(2):170–189, 1991.

Barbacci, M. R., "Instruction Set Processor Specifications (ISPS): The Notation and Applications," *IEEE Transactions on Computers,* C-30(1):24–40, 1981.

Bell, C. G., and A. Newell, *Computer Structures: Readings and Examples,* McGraw-Hill, New York, 1971.

Bergamaschi, Reinaldo A., "High-Level Synthesis in a Production Environment: Methodology and Algorithms," *Fundamentals and Standards in Hardware Description Languages—NATO ASI Series,* Kluwer Academic Publishers, Dordrecht, Netherlands, 1993, pp. 195–230.

Birman, Kenneth P., and T. A. Joseph, "Reliable Communication in the Presence of Failures," *ACM Transactions on Computer Systems,* 5(1):47–76, 1987.

Birman, Kenneth P., "The Process Group Approach to Reliable Distributed Computing," *Communications of the ACM,* 36(12):36–53, 1993.

Bochmann, Gregor V., "Specification of a Simplified Transport Protocol Using Different Formal Description Techniques," *Computer Networks and ISDN Systems,* 18, Elsevier Science Pub., 1990, pp. 335–377.

Boehm, Barry W., *Software Engineering Economics,* Prentice-Hall, New York, 1981.

Booch, Grady, *Object Oriented Design with Applications,* Benjamin/Cummings Pub. Co., Redwood City, Calif., 1991.

Borrione, Dominique, "CASCADE," *Fundamentals and Standards in Hardware Description Languages—NATO ASI Series,* Kluwer Academic Publishers, Netherlands, 1993, pp. 411–430.

CACI Inc., *SIMSCRIPT II.5 Programming Language,* CACI Inc., Los Angeles, 1987.

CCITT Recommendation Z.100, Functional Specification and Description Language SDL, AP IX-35, Geneva, 1988.

Clocksin, William F., and C. S. Mellish, *Programming in Prolog,* Springer-Verlag, Berlin, 1984.

Covington, Michael A., D. Nute, and A. Vellino, *Prolog Programming in Depth,* Scott, Foresman and Company, Glenview, Ill., 1988.

Cox, Brad J., and A. J. Novobilski, *Object Oriented Programming: An Evolutionary Approach,* Addison-Wesley, Reading, Mass., 1991.

Fishwick, Paul A., *Simulation Model Design and Execution—Building Digital Worlds,* Prentice-Hall, Englewood Cliffs, N.J., 1995.

Gajski, Daniel, et al., *High-Level Synthesis—Introduction to Chip and System Design,* Kluwer Academic Publishers, Norwell, Mass., 1992.

Gopalakrishnan, Ganesh C., "Formalization of an Operational Approach to Hardware Specification and Validation," *Proceedings of the 3rd Banff Higher Order Workshop,* Banff, Alberta, September 24–27, 1989, pp. 1–29.

Genrich, H. J., and K. Lautenbach, "System Modelling with High-Level Petri-Nets," *Theoretical Computer Science,* 13, North-Holland, Amsterdam, Netherlands, 1981, pp. 109–136.

Gronberg, P., et al., "PROD—A Pr/T-Net Reachability Analysis Tool," Helsinki University of Technology, Digital Systems Laboratory, Series B: Technical Reports, No. 11, June 1993.

Gupta, Aarti, "Formal Hardware Verification Methods: A Survey," *International Journal on Formal Methods in System Design,* 1(2/3):151–238, 1992.

Harel, David, "Statecharts: A Visual Formalism for Complex Systems," *Science of Computer Programming,* 8, Elsevier, Amsterdam, Netherlands, 1987, pp. 231–274.

Hayes, Robert H., et al., *Dynamic Manufacturing—Creating the Learning Organization,* Free Press, New York, 1988.

Hoare, C. A. R., *Communicating Sequential Processes,* Prentice-Hall, Englewood Cliffs, N.J., 1985.

International Organization for Standardization, Information Processing Systems, Open Systems Interconnection, LOTOS—A Formal Description Technique Based on the Temporal Ordering of Observational Behavior (ISO International Standard 8807), 1988.

Jacobson, Ivar, et al., *Object Oriented Software Engineering: A Use Case Driven Approach,* Addison-Wesley, Reading, Mass., 1992.

Jefferson, D. R., "Virtual Time," *Proceedings of the 1983 International Conference on Parallel Processing,* Calgary, Canada, August 1983, pp. 384–394.

Jensen, Kurt, "Coloured Petri-Nets and the Invariant Method," *Theoretical Computer Science,* 14, 317–336, Heidelberg, Germany, 1981.

Jensen, Kurt, "Coloured Petri-Nets," *Lecture Notes in Computer Science,* vol. 266, Springer-Verlag, Heidelberg, Germany, 1989.

Jerraya, Ahmed, P. G. Paulin, and S. Curry, "Meta VHDL for Higher Level Controller Modeling and Synthesis," *Proceedings of VLSI-91,* Edinburgh, August 1991.

Kleinrock, Leonard, *Queueing Systems,* vols. I and II, Wiley, New York, 1976.

Klick, Vickie B., J. Patti, and M. L. Todd, "Experiences in the Use of SDL/GR in the Software Development Process," in O. Faergemand and R. Reed (eds.), *SDL '91 Evolving Methods,* Elsevier, Amsterdam, Netherlands, 1991, pp. 449–457.

Koomen, C. J., *The Design of Communicating Systems: A System Engineering Approach,* Kluwer Academic Publishers, Boston, 1991.

Kumar, A., V. Kashyap, S. D. Sherlekar, G. Venkatesh, S. Biswas, P. C. P. Bhatt, and S. Kumar, "IDEAS: A Tool for VLSI CAD," *IEEE Design and Test of Computers,* 6(5):50–57, 1989.

Lakos, C. A., "LOOPN—Language for Object-Oriented Petri Nets," Technical Report 91-1, Department of Computer Science, University of Tasmania, 1991.

Lanneer, Dirk, et al., "Architectural Synthesis for Medium and High Throughput Signal Processing with the New CATHEDRAL Environment," in R. Camposano and W. Wolf (eds.), *High-Level VLSI Synthesis,* Kluwer Academic Publishers, Norwell, Mass., 1991, pp. 27–54.

Lewis, Ted., *Software Engineering—Analysis and Verification,* Reston, Va., 1982.

McFarland, Michael C., A. C. Parker, and R. Camposano, "The High-Level Synthesis of Digital Systems," *Proceedings of the IEEE,* 78(2):301–318, 1990.

Milner, R., "A Calculus of Communication Systems," *Lecture Notes in Computer Science,* Vol. 92, Springer-Verlag, London, 1980.

Moore, Andrew P., "The Specification and Verified Decomposition of System Requirements Using CSP," *IEEE Transactions on Software Engineering,* 16(9):932–948, 1990.

Morrison, John D., and C. O. Newton, "ELLA," *Fundamentals and Standards in Hardware Description Languages—NATO ASI Series,* Kluwer Academic Publishers, Dordrecht, Netherlands, 1993, pp. 385–394.

Moszowski, Ben, "A Temporal Logic for Multilevel Reasoning about Hardware," *IEEE Computer Magazine,* February 1985, pp. 10–19.

Murata, Tadao, "Petri Nets: Properties, Analysis and Applications," *Proceedings of the IEEE,* 77(4):541–580, 1989.

Parent, Pierre, and O. Tanir, "Voltaire: A Discrete Event Simulator," *Proceeding of the PNPM-91 Workshop (Tools Amendment),* Melbourne, December 1991.

Peterson, James L., *Petri Net Theory and the Modeling of Systems,* Prentice-Hall, Englewood Cliffs, N.J., 1981.

Petri, C., "Kommunikation mit Automaten (Communication with Automata)," Ph.D. dissertation, University of Bonn, Bonn, Germany, 1962.

Piloty, R., et al., "An Overview of Conlan," *Proceedings of the IFIP Congress,* Tokyo, 1980.

Pritsker, A. A., *Introduction to Simulation and SLAM II,* Halsted Press, West Lafayette, Ind., 1986.

de Prycker, Martin, *Asynchronous Transfer Mode—Solution for Broadband ISDN,* Ellis Horwood Ltd., London, 1991.

Ramming, Franz J., "System Level Design," *Fundamentals and Standards in Hardware Description Languages—NATO ASI Series,* Kluwer Academic Publishers, Dordrecht, Netherlands, 1993, pp. 109–151.

Rattray, C. (ed.), *Specification and Verification of Concurrent Systems,* Springer-Verlag, New York, 1989.

Reed, Daniel A., and R. M. Fujimoto, *Multicomputer Networks—Message Based Parallel Processing,* MIT Press, Cambridge, Mass., 1987.

Reisig, Wolfgang, "Petri-Nets—An Introduction," EATCS Monographs on Theoretical Computer Science, Springer-Verlag, New York, 1982.

Ross, Sheldon M., *Introduction to Probability Models,* Academic Press, San Diego, 1992.

Saultz, J. E., RASSP Final Technical Report (CLIN 0002AB), CMDA972-92-R-0017, GE Aerospace—Advanced Technology Laboratories, Moorestown, N.J., Oct. 21, 1992.

Schriber, Thomas, *Simulation Using GPSS,* Wiley, New York, 1974.

Shlaer, Sally, and S. Mellor, *Object Life-Cycles—Modelling the World in States,* Prentice-Hall, Englewood Cliffs, N.J., 1992.

Srivas, Mandayam, and M. Bickford, "Formal Verification of a Pipelined Microprocessor," *IEEE Software Magazine,* September 1990, pp. 52–64.

Tanir, O., V. K. Agarwal, and P. C. P. Bhatt, "A System Level Synthesis Framework for Computer Architecture," *Proceedings of the Third International Workshop on Rapid System Prototyping,* Research Triangle Park, N.C., June 23–25, 1992, pp. 94–111.

Tanir, O., V. K. Agarwal, and P. C. P. Bhatt, "The Design of a Library Support System for a Telecommunication System Synthesis Environment," *Proceedings of the Fourth International Workshop on Rapid System Prototyping,* Research Triangle Park, N.C., June 28–30, 1991, pp. 54–67.

Tanir, Oryal, V. K. Agarwal, and P. C. P. Bhatt, "On the Design of Real-Time Telecommunication Systems," *Proceedings of the IEEE Workshop on Real-Time Applications,* New York, May 11–12, 1992.

Tanir, Oryal, V. K. Agarwal, and P. C. P. Bhatt, "Architectural Telecommunication Modelling," presented at the Co-Design Workshop, Cambridge, Mass., Oct. 7, 1993a.

Tanir, Oryal, V. K. Agarwal, and P. C. P. Bhatt, "System Design Exploration in a Specification Driven Simulation Environment," *Proceedings of the European Simulation Symposium,* SCS International, Delft, Netherlands, Oct. 25–28, 1993b, pp. 591–596.

Tanir, Oryal, V. K. Agarwal, and P. C. P. Bhatt, "DASE: An Architectural Level System Design and Modelling Environment," *Proceedings of the International Conference on Concurrent Engineering and Electronic Design Automation,* SCS International, Bournemouth, U.K., April 7–8, 1994a, pp. 414–421.

Tanir, Oryal, V. K. Agarwal, and P. C. P. Bhatt, "DASE: An Environment for System Level Telecommunication Design Exploration and Modelling," *Fourth International CAST 94 Workshop,* Ottawa, Canada, May 16, 1994b; proceedings to be published by Springer-Verlag.

Tanir, Oryal, V. K. Agarwal, and P. C. P. Bhatt, "A Specification Driven Architectural Design Environment," *IEEE Computer Magazine,* 28(6):26–35, 1995.

Tanir, Oryal, and S. Sevinc, "Defining the Requirements for a Standard Simulation Environment," *IEEE Computer Magazine,* February 1994c, pp. 28–34.

Thomas, D. E., and T. E. Fuhrrman, "Industrial Uses of the System Architect's Workbench," in R. Camposano and W. Wolf (eds.), *High-Level VLSI Synthesis,* Kluwer Academic Publishers, Norwell, Mass., 1991, pp. 307–330.

Trillium Rel. 3.01, Model for Telecom Product Development and Support Process Capability, June 1995, Bell Canada.

Verilog Hardware Description Language Reference Manual, Open Verilog International, 1991.

IEEE Standard VHDL Language Reference Manual, IEEE IStd 1076, 1987.

Ward, Paul T., and S. J. Mellor, *Structured Development for Real-Time Systems,* Vol. 1, Yourdon Press Computing Series, Prentice-Hall, Englewood Cliffs, N.J., 1985.

Wernik, Marek, and E. A. Munter, "Broadband Public Network and Switch Architecture," *IEEE Communications Magazine,* January 1991, pp. 83–89.

Zeigler, Bernard P., *Object Oriented Simulation and Hierarchical Modular Models—Intelligent Agents and Endomorphic Systems,* Academic Press, San Diego, 1993.

Zeigler, Bernard P., *Multifaceted Modelling and Discrete Event Simulation,* Academic Press, Orlando, Fla., 1984.

Zegura, Ellen Witte, "Architecture for ATM Switching Systems," *IEEE Communications Magazine,* February 1993, pp. 28–37.

Index

ABOUT THE AUTHOR

Oryal Tanir is a Senior Consultant with Bell Canada in the field of broadband network risk management, and simulation, and system design. He is also an adjunct professor at McGill University. Dr. Tanir initiated and chaired the IEEE Working Group on Simulation and has participated in organizing committees for numerous simulation conferences throughout the world.